Fort Collins' *First* Water Works

1882-1883 Fort Collins Water Works pump house, ca. 1900-1910. Courtesy, Denver Public Library, Western History Collection, L. C. McClure, MCC-1066. The cover photograph is the same L. C. McClure image of the pump house.

Fort Collins' *First* Water Works Again a Work in Progress

by Wayne C. Sundberg

with contributions by
Albert Barnes, David Budge, Richard Carrillo,
Kevin Cook, Norman Evans, Philip Hoefer, Susan Quinnell, and Brian Werner

Edited by Susan Hoskinson

Poudre Landmarks Foundation, Inc.
Fort Collins, Colorado

Poudre Landmarks Foundation, Inc., Fort Collins, Colorado 80521

Printed in the United States of America.

Library of Congress Control Number 2004092656

ISBN 0-9753849-3-7

This project was partially funded by a State Historical Fund grant award from the Colorado Historical Society.

Cover and frontispiece photograph: 1882-1883 Fort Collins Water Works pump house, ca. 1900-1910. On assignment to photograph the new Greeley, Colorado, water filtration system near the Poudre Canyon entrance, L. C. McClure stopped briefly at the Fort Collins Water Works to capture the earliest known image of the pump house. This photograph recorded improvements to Fort Collins' water delivery system made in 1894, including construction of a stone wing wall behind the building to enlarge the reservoir and the addition of a second room to house the auxiliary steam pump. It shows a portion of the stone arch over the tail race, stone wing wall, drop structure, second addition, and smoke stack held in place by guy wires. Courtesy, Denver Public Library, Western History Collection, L. C. McClure, MCC-1066.

Dedication

This book is dedicated to Jim and Doris Greenacre, who were the driving force behind the preservation of the first City of Fort Collins Water Works. Their son aptly described their zeal for historic preservation. Jim Greenacre said about his parents: "She [Doris] was a woman who felt it was her duty, as did my father, to speak out and change what they could so that part of the 'Magic of Fort Collins' would be preserved for those of us who are left behind."

Jim and Doris returned to their hometown, Fort Collins, Colorado, upon retirement in 1972. This is where family roots ran deep to great-grandparents, grandparents, and parents. When they set themselves on a mission to preserve the 1882-1883 Fort Collins Water Works, they found allies in Mike Smith of the City of Fort Collins Water Utilities Department and the Fort Collins Water Board. Beginning in 1986, Doris and Jim offered persistent and convincing arguments for preserving the more than twenty-six acre property and its buildings. Although in 1971 the decision had been made to designate the pump house as a local landmark, seventeen years elapsed before funds were budgeted for structural stabilization of the historic brick and stone building.

Seven years later, in 1995, the site was leased to the Poudre Landmarks Foundation by the City of Fort Collins with the provision that it be preserved as a local historic district, and the Water Works building be used to interpret the water resources of the city and Cache la Poudre River basin.

That the Water Works property, a significant facility in the heritage of Fort Collins, is being preserved today is due to the devotion and commitment of Jim and Doris Greenacre.

Norman Evans,
Poudre Landmarks Foundation
and Friends of the Water Works

Contents

List of Photographs and Maps

PHOTOGRAPHS

MAPS

Foreword

The 1882-1883 Fort Collins Water Works was constructed to protect public health and safety arising out of the need to fight fire and provide clean drinking water for the community. The citizens of Fort Collins recognized the need for fire protection in the early 1880s after experiencing several disastrous fires. One even resulted in the loss of life.

The need was not unique to Fort Collins, however. Since the earliest days of colonial America and territorial Colorado, fire disasters greatly impacted our history. Part of the problem was the way people lived. Wood was used to construct everything from houses and barns, to sailing ships, bridges, and railroad cars. Open flames were used for cooking, heating, and lighting.

New construction techniques using brick, nonflammable roofs, and metal window shutters provided some protection. What was needed was a way to fight fire, one-on-one, with pressurized water. Today, fire hoses deliver one thousand gallons of water per minute. It took time, understanding, and good equipment to get to this point.

At the same time fire was running rampant in America, so were diseases associated with contaminated water. Typhoid fever, dysentery, and cholera have plagued humanity throughout time. The risk of infection was greater in high-density populations, such as those found in nineteenth-century industrial towns, usually in the summertime. The germ theory of disease is generally attributed to French chemist Louis Pasteur, who by the 1850s had shown that microbes, and not chemistry, caused fermentation. Until the relationship between germs and diseases was proven, humanity suffered greatly.

But as doctors discovered the mechanism of disease and the dangers of dirt, and engineers pioneered mass plumbing, the world began to change. A 1916 study showed how disease could be successfully controlled through water supply improvements. Annual typhoid death rates for various American cities per hundred thousand population between 1880 and 1914 show significant decreases in disease frequency as a result of the improvements that communities made in their water supply systems. Fort Collins was no exception.

Today, the existence, location, and decline of the1882-1883 Fort Collins Water Works provide valuable historical insight into the environmental and human events that preceded its construction, operation, and use after the facility was decommissioned in 1905. As twentieth-century technology advanced, its importance waned.

History offers a profusion of general and specialized surveys and monographs about westward expansion. Few books deal with the small-scale, publicly constructed facilities that improved everyday life for growing communities. They also have an important story to tell.

This book is an attempt to provide readers with an understanding of one of the major accomplishments of the people who selected Fort Collins as their home and the changes they brought to the area. It also tries to explain the technological triumphs that made settlement of the area possible.

David Budge,
Water Works Project Manager

Township Description, 1864

Township 7 N R 69 W has a quantity of good farming land along the cache la pouder [sic] Tp 8 N R 69 W also has some good farming land. The town of LaPorte [sic] is situated in this tp. This tp should be subdivided.

General Land Office survey field notes of the exterior lines of Township 6, 7, and 8 North, Range 69 West of the Sixth Principal Meridian between October 15 and October 20, 1864.

In 1882-1883, the City of Fort Collins located its first water pumping plant in the south half of Section 32, Township 8.

--Handwritten field notes as transcribed by Albert H. Barnes,
in "Land Surveying for the Fort Collins Waterworks"
(Unpublished paper, Fort Collins Water Works Archive, 2003).

EDITOR'S NOTE: Although the area's first surveyors referred to this early Colorado community as LaPorte, current practice refers to the unincorporated town as Laporte. In the French language, "la porte" means "gateway."

1882-1883 Fort Collins Water Works pump house, 1974. Water from Larimer County No. 2 Canal pours over the drop structure, located south of the pump house, into the channel that leads to the Cache la Poudre River. Notice the increased amount of vegetation around the pump house compared with the 1900-1910, L. C. McClure photograph on the cover and frontispiece. Courtesy, Fort Collins Public Library, Triangle Review collection.

Introduction

On the corner of Overland Trail and Bingham Hill Road, south of Laporte, Colorado, near the Cache la Poudre River, stand a few buildings on the edge of a large pasture. Tall cottonwood trees form a lush backdrop in summer. A quick glance as you drive by reveals a long, brick structure, a frame house, and some outbuildings. You think, "someone should do something with that property."

They are.

This book is not only a history of the 1882-1883 Fort Collins Water Works but also a report to the community about the progress made during the past ten years by the Friends of the Water Works, an auxiliary of the Poudre Landmarks Foundation, as they develop the property into an interpretive center. Volunteers and a few paid professionals are delving into its history, locating archaeological evidence, and compiling architectural, ecological, and agricultural reports for later use. Their work will become the basis for interpreters to create exhibits and programs that will help the public better understand the precious nature of water, especially in the arid West. The more than twenty-six acre property offers many interpretive opportunities, but the focus of the center will be on water as it affected the growth of Fort Collins and its surrounding agricultural community.

The 1882-1883 Fort Collins Water Works, therefore, is a work in progress undertaken by the Friends of the Water Works. Prior to Foundation involvement in 1994, important building stabilization measures halted the slow deterioration of this unique example of nineteenth-century industrial architecture. At the urging of Doris and Jim Greenacre, who brought public attention to the property's sad plight in 1986, the City of Fort Collins Water Utilities Department underwrote a project to prevent precipitation, debris, and vandals from causing further damage to the city's first water supply facility.

Stabilization allowed the ad hoc committee that formed around the Greenacres' preservation effort to begin planning possible uses for the property. A series of meetings with city and county officials, and interested individuals and groups explored how the long-abandoned building could be used. Owned by the City of Fort Collins and now leased by the Poudre Landmarks Foundation, the parcel includes the original brick pump house with two brick additions. One addition housed a steam-powered pump, and the other served as a filter room. A caretaker's house with associated outbuildings and apple orchard, plus two parallel irrigation canals also are prominent features of the property. Its nearness to the Cache la Poudre River and the abundance and diversity of vegetation suggest that those who visit the property could learn about its natural as well as its municipal and industrial histories. All who attended the early meetings about the property agreed that the goal should be to create an interpretive center to tell about water development and its impact on the area.

Once the Foundation's Board of Directors formally adopted the project in 1995, concern about possible vandalism led the Friends group to entertain a proposal by the Rocky Mountain Raptor Program, a non-profit organization associated with Colorado State University, to occupy the caretaker's quarters. One of the first steps toward achieving security for the property was to renovate the house, again with financial assistance from the city's water department. Completed under the direction of Friends volunteer Loren Maxey in 1996, the house has been occupied ever since by individuals who have participated in the Raptor Program. During Water Works open houses, Raptor Program members display injured birds of prey that have been healed.

Ten years ago, the Friends also undertook the tasks of conducting research about the property and site planning. On behalf of the Friends in 1995, Sally Ketcham wrote a successful grant proposal to the State Historical Fund to prepare a preservation and site-development plan. Merrick and Company completed the plan in 1997, in consultation with Friends volunteers Jane Hail, Colorado State University department of history professor John Albright, and other Poudre Landmarks Foundation Board members. With a plan to follow, building restoration became the focus. Colorado Questers and the Questers International organizations supported repair of the facility's windows and doors, completed in 1999.

In the meantime, local historians Wayne Sundberg, David Budge, Jean Petersen, and other Friends formed a Water Works History Committee. They began the search for documents, photographs, maps, and oral histories to answer elusive questions about pump house construction and operation, and the overall water delivery system to town. On contract with the Friends, Sandy Dion collected and organized copies of old newspaper articles in notebooks that became a primary source for this book. Minutes from early meetings of the town's Board of Trustees, utility department reports, and the newspaper articles also offered insight into the attempts made by a small community to address its need for a reliable water system. *Fort Collins Courier* editor Ansel Watrous published a general summary of how the Water Works was constructed and operated in 1883, and until the Friends' Water Works project expanded to include pump house excavations as a means to learn more about the property, newspaper accounts have been the main source for information about the plant's inception, construction, improvements, and eventual demise.

As more people became interested in the picturesque property, they encouraged the Friends to conduct various other surveys to record existing natural resources. Water Works Project Manager David Budge enlisted the help of various specialists to enable the Friends to better develop the site as an ecological resource. David continues to devote countless hours to all facets of the Water Works project. His professional experience as a historical geologist and geological engineer enabled the Friends to undertake not only building restoration but also landscape revegetation and vitalization through weed control.

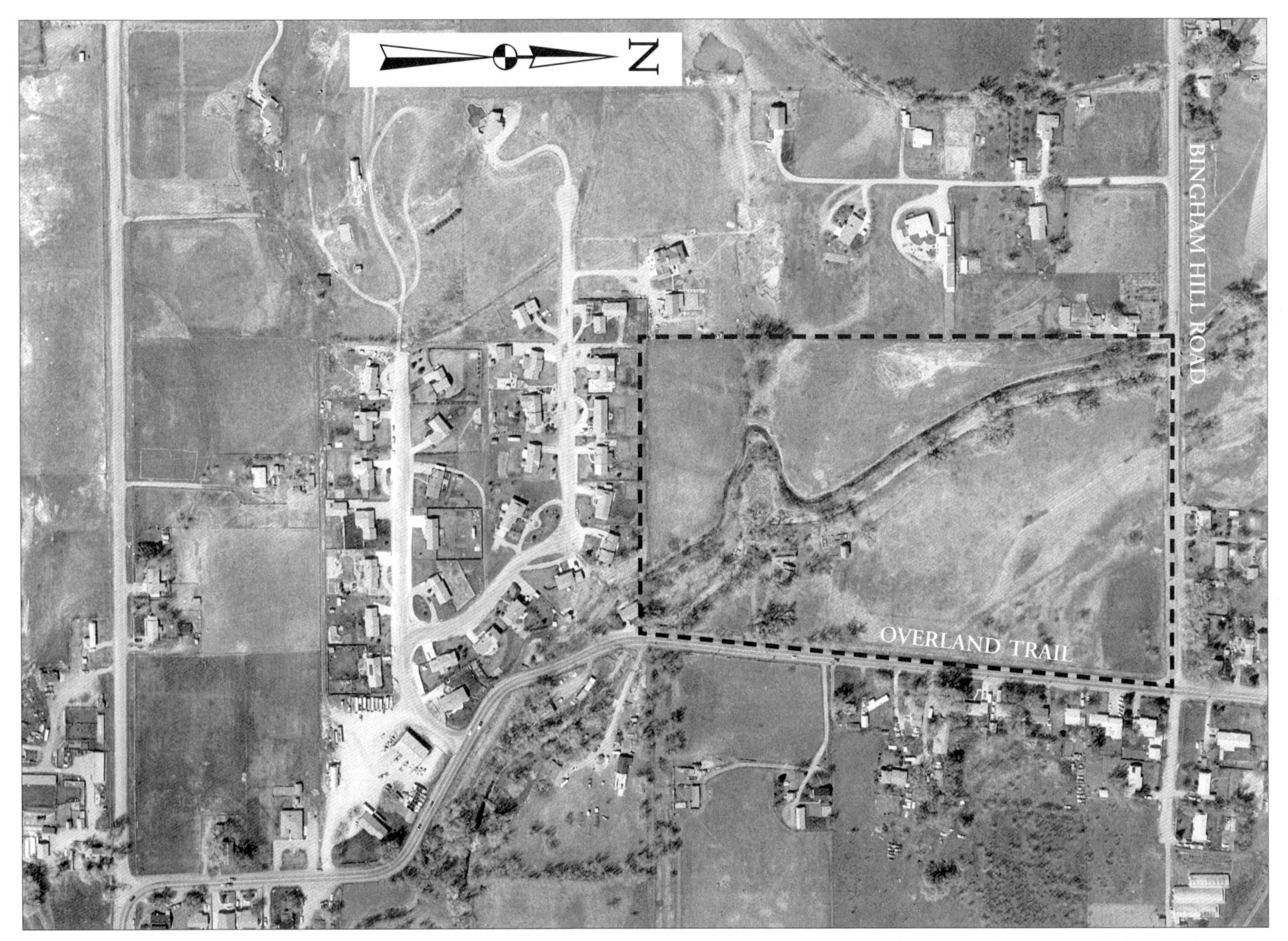

Aerial photograph of Fort Collins Water Works property, 1987. Bounded on the east by Overland Trail and on the north by Bingham Hill Road, two irrigation canals cross the property diagonally. Neighboring properties include a housing subdivision and residential acreages. Courtesy, Fort Collins Water Works Archive. Photography by Aero-Metric, Inc.

Taking the first step toward creating the natural area, retired Colorado State University range science professor Clinton Wasser identified two hundred plant species growing on the property, of which twenty percent are native to the area. Then, a team from the U.S. Department of Agriculture Natural Resources Conservation Service presented a long-term management plan for weed control, adopted by the Friends in 1999, in order to re-establish native grasses on the property. To carry out this plan, the Larimer County Weed Control District employees now work closely with the Friends to control noxious weeds with fertilizer, sprays, and biological means. In addition, several small test plots of native grasses are being monitored from year to year.

Before the ground could be plowed and seeded to establish these test plots, however, it was necessary to survey the pasture for archaeological resources. Work began in 1998 when Colorado State University anthropology professor Calvin Jennings and several students excavated test pits near a long-abandoned road right-of-way that runs through the property. One year later, several members from the Rocky Mountain Prospectors and Treasure Hunters Club used metal-detectors to survey one and one-half acres in the northeast corner where the native grass test plots would be planted. Surveyors found 389 artifacts, and Colorado State University anthropology professor Mary Van Buren, who was in charge of the survey, concluded that a variety of activities had taken place on the site from the late nineteenth century to the present.

A study of the trees on the property began in 1997 when former Colorado State Forest Service forester Philip Hoefer discovered that the cottonwood standing at the property's entrance qualified to be among one of the largest known in Larimer County. With a circumference of 310 inches at four and one-half feet above the ground, the one hundred foot tall tree could be more than 120 years old. Seepage from the two irrigation ditches that cross the acreage help the cottonwoods thrive, and also keep seven varieties of apple trees alive in the hundred-year-old orchard. Scott Skogerboe, a local nurseryman, grafted cuttings from the old trees onto new rootstock. Seven of the "new" trees now

Trees were not a dominant landscape feature on the plains of Colorado prior to European settlement. Minimal rainfall kept trees from naturally growing, except along rivers, streams, and other wetlands. Most trees require a minimum of twenty inches of natural moisture. The average annual rainfall in this area is fifteen inches.

The 1882-1883 Fort Collins Water Works is situated on the western edge of this high plains desert. It is not too difficult to determine what trees or shrubs grew in this area. All one needs to do is go north of Fort Collins and Laporte and observe areas of uncultivated lands abutting the foothills. Most likely no woody vegetation was growing at the Water Works site prior to settlement and development. A critical factor for tree growth on the high plains is moisture.

European settlement introduced water diversion. With the benefit of moisture, native trees eventually seeded and rooted along contoured water corridors. The key tree to begin following these new waterways was the native plains cottonwood (Populus sargentii). Other species that would follow were boxelder (Acer negundo), willow (Salix spp.), and alder (Alnus tenuifolia). After the 1882-1883 Fort Collins Water Works established a water source behind the pump house, these same native trees began growing on the property.

--Philip Hoefer,
"Trees of the Old Water Works"
(Unpublished paper, Fort Collins
Water Works Archive, 2002).

stand next to their "parents" and will preserve the orchard. Later, Colorado State University horticulture professor Cecil Stushnoff identified the apple varieties, some no longer actively raised.

Although water seepage from the canals encouraged growth of trees and grasses, it caused structural damage to the original pump house and the spillway, or drop structure, next to the building. The Friends reached this conclusion after David Budge noted evidence that the structure was settling. Volunteer Norman Evans, a former City of Fort Collins Water Board member and retired Colorado State University engineering professor, then began monitoring the seepage and recording water table data for the past several years before repairs were undertaken recently.

A project to stabilize the pump house walls began in 2003 with State Historical Fund support. In preparation, historical archaeologist Richard Carrillo of Cuartelejo HP Associates, with assistance from dozens of volunteers, excavated portions of the pump house interior and exterior beginning in 2000, work also underwritten by a State Historical

Deer and other wildlife still find refuge at the park-like Water Works acreage. Courtesy, Lisa Steffes.

Fund grant. These ongoing excavations allow student and community volunteers to gain "hands-on" experience in historical archaeology. To date, stone piers that held the pumps, a stone archway beneath the building, and wrought iron pipes visually confirm much of the early newspaper descriptions of how the Water Works really worked.

What no one expected when work first began on the Water Works property was the discovery of wagon ruts from an old roadway leading to the river crossing at Laporte. Identifying those ruts involved the work of surveyors, geologists, aerial photographers, archaeologists, and historians studying old maps and diaries. Explaining the presence of ruts adds to the excitement surrounding the property because they may be evidence of one leg of the old Overland Stage route, also called the "Cherokee Trail."

In 1998, Larimer County surveyors re-located an abandoned road that once crossed the Water Works property by comparing Larimer County maps from the 1870s and 1880s. The puzzle was coming together. Archaeologists found rusted

Artifacts from archaeological survey. Volunteers and archaeologists located 389 industrial-type artifacts during a 1999 metal detector survey. Two objects, the ringed ferrule (No. 33) and wedge (No. 177), demonstrate the variety of activities that took place on the property. The rusted ferrule may be related to the old wagon road identified as one leg of the Overland Stage route, also called the "Cherokee Trail." The wedge may have been used for splitting logs. Courtesy, David Budge.

wagon hardware during the 1999 surveys and concluded that the parts could be related to the old stage route that crossed the property. In 2001, a ground-penetrating radar study in the same general area noted two depressions that also indicated remnants of a route or trail. During the 2002 excavations, historical archaeologist Richard Carrillo obtained soil profiles and strengthened the argument that the depressions probably are a portion of the 1850s Cherokee Trail, once traveled by Horace Greeley on his way to San Francisco in 1859.

So the research continues. Questions answered lead to more that are unanswered. We hope that through the pages of this book you will answer some of your own questions about the Water Works property. The book should give you a better understanding of the hurdles early Fort Collins' community leaders surmounted in order to provide the town with its first water system. And, the Friends of the Water Works and the Poudre Landmarks Foundation want the interpretive center to soon become a reality. It will demonstrate how the wise use of water resources allowed a small town such as Fort Collins to become a dynamic, thriving modern city.

Thanks to the many volunteers not mentioned by name, the continued support of the City of Fort Collins and its Water Utilities Department, and the State Historical Fund, a division of the Colorado Historical Society, for making the Water Works Project and this book possible.

Susan Hoskinson
Editor

Cache la Poudre River in Pleasant Valley, ca. 1900-1910. The river offered a pure, more consistent source of water for household and commercial uses to Fort Collins' earliest residents. As more people moved to the area, demand for water increased. This vantage point is several miles northwest of the original Fort Collins Water Works diversion structure. Courtesy, Denver Public Library, Western History Collection, L. C. McClure, MCC-1069.

Chapter 1: Before Fort Collins Became a City

For more than ten thousand years, the Cache la Poudre River valley has been a gathering place for people. Various Native American groups found the valley to have an abundant supply of food and water. People of the Folsom culture hunted giant prehistoric *bison antiqus* on the high plains north of the river and left archaeological evidence of their stay in the soil of the Cheyenne Ridge. Later, Cheyenne and Arapaho hunters pursued the smaller American bison, pronghorn, and deer along the Poudre and its tributaries.

Trappers, traders, and travelers of the nineteenth century followed the river as they traversed this part of the West. Wild animals and birds used the river as a source of life-giving water and provided meat for these transients who passed along its course. Some of these early passersby even left the melodic name, Cache la Poudre, "hiding place of the powder," on the river in the early 1800s. Another left us one of the earliest descriptions of the river. Antoine Janis, who later became one of the valley's first settlers, passed through in 1844 on a trading foray out of Fort Laramie. Climbing to the top of a hill, perhaps present-day Bingham Hill, he viewed the valley that would later be his home. In a letter he wrote to Ansel Watrous in March 1883, Janis recalled:

> On the first of June, 1844, I stuck my stake on a claim in the valley,...At that time the streams were all very high and the valley black with buffalo. As far as the eye could reach, nothing scarcely could be seen but buffalo. I was just returning

Standing at the 1882–1883 Fort Collins Water Works today, one can hardly imagine dinosaurs grazing the lawn. As time passed, the landscape changed. And, the 1882–1883 Fort Collins Water Works site continues to evolve as a wildlife community.

—Kevin Cook,
"Wildlife of the 1882-1883 Fort Collins Water Works"
(Unpublished paper, Fort Collins Water Works Archive, 2002).

> from [New] Mexico* and I thought the Poudre valley was the loveliest spot on earth, and think so yet.[1]

In 1849 and 1850, part-Cherokee miners from Georgia followed the river on their journey to and from the California goldfields. The Overland Stage road passed through the Fort Collins Military Post and Laporte, both located to tap the river's water supply. One of the early travelers left a vivid description of the Poudre River. J. R. Todd traveled upstream from the South Platte River in June 1852. He later wrote:

> The waters of the river were as clear as crystal all the way down to its confluence with the Platte. Its banks were fringed with timber,...consisting of cottonwood, boxelder, and some willow. Its waters were full of trout of the speckled or mountain variety. The undulating bluffs sloped gently to the valley which was

*Janis probably was returning from Taos. In 1844, Taos was part of Mexico, and the Arkansas River divided Mexico from the territory claimed by the United States.

> carpeted with the most luxuriant grasses.... Wild flowers of the richest hue beautified the landscape, while above all towered the majestic Rocky Mountains to the westward of the valley....[2]

The river and its water defined the area's attraction for people dependent on agriculture.

Wells and Ditch Water Supplied the Town

So what were the earliest sources of water as soldiers and settlers began to inhabit the Cache la Poudre valley? Both Camp Collins and Fort Collins were built on the banks of the river. This proved to be fatal to Camp Collins at its location above Laporte when too much water appeared in the form of a flood in June 1864. A replacement fort, Fort Collins, was built five miles downstream, on the higher south side of the river. Watering pens for the horses were located below the high bank of the river, and drinking water for the soldiers was undoubtedly drawn from the river above these corrals. One soldier spoke of hunting ducks at a pond south of the fort. A later sketch map of the town shows a

[In 1842], Captain John C. Fremont's party reached Fort St. Vrain on the South Platte near the mouth of the St. Vrain River and turned north toward the Big Thompson and Cache la Poudre rivers. Fremont described this northward leg as "fields of flowers resembling a garden." The Poudre River was a beautiful mountain stream, about one hundred feet wide, flowing swiftly over a rocky bed. The party proceeded upstream to the mouth of the river where they turned north again toward Fort Laramie.

—Norman Evans,
"Colorado Irrigation History—
The Cache la Poudre River Basin"
(Unpublished paper, Fort Collins Water
Works Archive, 2002).

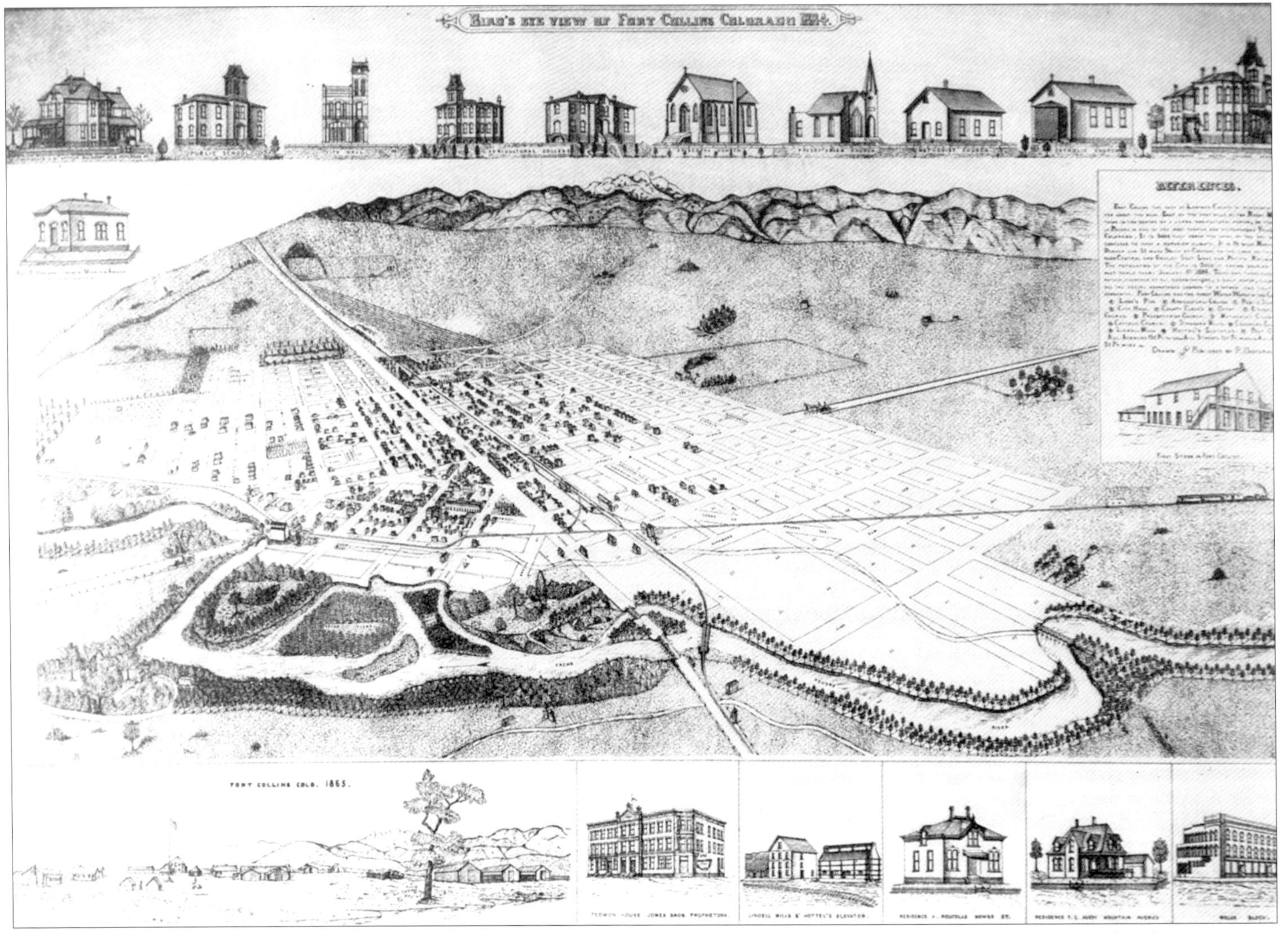

"Bird's Eye View of Fort Collins, Colorado, 1884." Mapmaker Pierre Dastarac located a pond near College Avenue and Lake Street (upper left hand corner of map). College Avenue is the widest thoroughfare. The Cache la Poudre River meanders at the bottom of the map. Courtesy, Fort Collins Public Library.

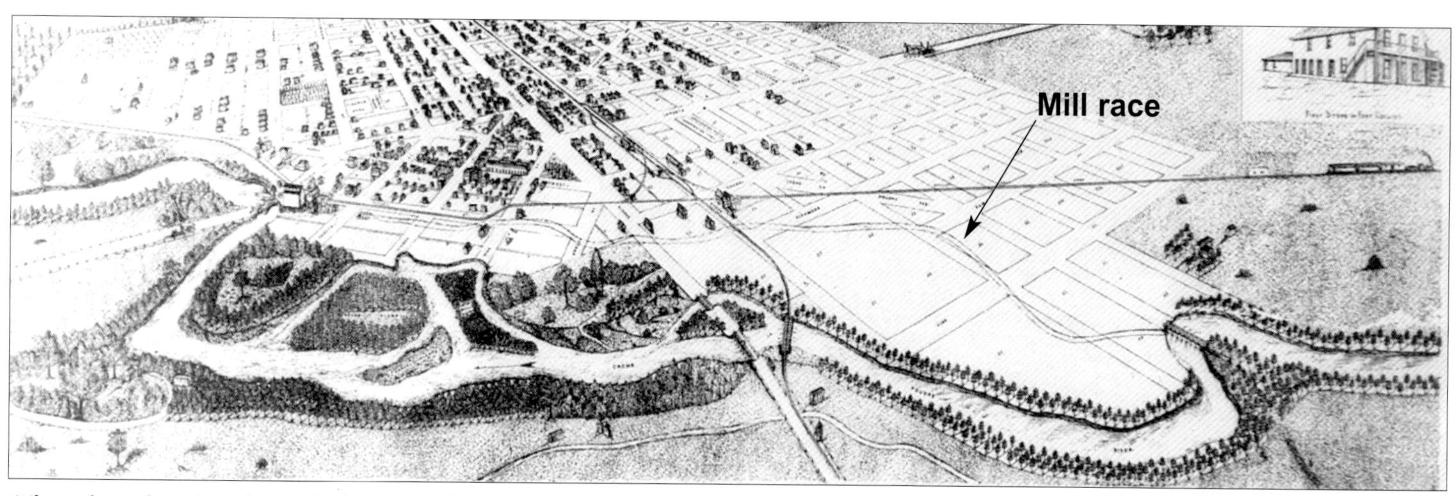

The enlarged section shows the location of the mill race, or channel, that brought water from the Cache la Poudre River to the Mason and Hottel mill. Today, the mill structure is part of Ranch-Way Feed Mills. Courtesy, Fort Collins Public Library.

small lake or pond on the east side of South College Avenue where Lake Street intersects College. Early residents also reported a swale, or marshy low-lying area, near the intersection of Remington and Elizabeth streets. Some residences probably had wells in their yards. The fort and early town seemed to have had several sources of water, even though there was no town water system.

Prior to building the two military camps, farmers began digging a system of irrigation canals to water the river's bench lands. These ditches later provided another possible source of domestic water. The "mill race," or channel, dug to carry water to the mill of "Auntie" Elizabeth Stone and Henry Clay Peterson in the late 1860s, took water from the south side of the river a little north of the fledgling town. It also provided another potential source of town water. In addition, potable water was sold from "water wagons" at a cost of five cents a bucket and twenty-five cents a barrel. Local entrepreneur

Fort Collins water wagon, n.d. Water may have been delivered door-to-door in the 1880s with a wagon like this. Later, the outfit may have served as a sprinkler wagon to control dust on roads during construction. Courtesy, Fort Collins Public Library.

John Peterson placed an advertisement dated August 10, 1880, in the August 19, 1880, edition of the *Fort Collins Weekly Express:*

> I have secured a complete outfit and will begin Thursday, August 12th, to deliver water to any part of the city. I shall take water directly from the river [Poudre], and will take special pains for it to be pure and wholesome. I would respectfully solicit the patronage of those wishing water for domestic purposes.

All these sources more or less met the water needs for most homes and businesses, but they did not meet all the water requirements of a growing community.

Two principal factors influenced the development of irrigation in the nineteenth and early twentieth centuries in Northern Colorado: the region's semi-aridity and Colorado's peculiar geography that leads to dependence on mountain snows to provide the majority of our water supply.

—Brian Werner,
"Irrigation Development in Northern Colorado:
A Brief History of How Water Influenced
the Development of the Fort Collins Region"
(Unpublished paper, Fort Collins Water
Works Archive, 2002).

Linden and Willow streets, 1881. Although most early Fort Collins stores and businesses were built of wood, the threat of devastating fires encouraged many to rebuild in brick. By 1881, wood-frame and brick commercial buildings stood side-by-side. Looking southwest, the imposing three-story brick structure on the corner of Linden and Jefferson streets (right center) is the Tedmon House, built of brick in 1880. With a bathroom on each floor, it then was considered a luxury hotel. Old Grout stands across the street, soon to make way for Frank Stover's Keystone Block, scene of the devastating 1882 fire. Courtesy, Fort Collins Public Library.

Chapter 2: Town Needs a New Water System

Most of the earliest homes and businesses were constructed of wood after the town was officially surveyed and platted in early 1873; consequently, fires were a major concern. Lacking a municipal water system, the "bucket brigade" was the main line of defense against fires. By the late 1870s, more structures were built of brick and stone, somewhat "fireproofing" them. By the early 1880s, three-story buildings, such as the Tedmon House at Linden and Jefferson streets, the Opera House Block on College Avenue, and numerous two-story businesses along Jefferson Street, gave the town a more permanent air. As their heights soared to two and three stories, throwing buckets of water on them was an ineffective defense against consumption by a fiery conflagration.

Ansel Watrous, in *History of Larimer County*, pointed out, that in addition to too many saloons, houses of prostitution, and gambling halls, another problem was fires:

> To make matters still worse, incendiarism was rampant, hardly a night occurring without one or more alarms of fires. The loss in some of these fires was heavy, while in other instances only sheds, stables and barns were burned. The town was full of idle and vicious men, driftwood from railroad and ditch camps, irresponsible creatures, without home or friends, who hung about the saloons and brothels. Several of the fires were laid at the door of these men and were started for the sole purpose of getting a free drink. They had noticed that whenever a fire occurred, the saloons set up a drink to the fire fighters after they had gotten the flames under control....[3]

Wooden storefront, ca. 1905. Moved to West Mountain Avenue in 1880, the first wooden commercial structure built on College Avenue housed the law offices of Jay H. Boughton and the offices of the Fort Collins Standard. *Clark Boughton was editor of the* Standard, *Fort Collins' second newspaper. The 1874 structure was moved from the 100 block of North College Avenue around the corner to West Mountain to make way for the Opera House Block. It was razed in 1905 to make way for the Barkley, Boughton, and Crain Hardware store. Courtesy, Fort Collins Public Library and Wayne Sundberg.*

The anti-saloon element and the pro-water system people were beginning to have a rallying point, with more reasons to come.

Jacob Welch built a large, two-story wooden dry goods store at the northwest corner of Mountain and College avenues in 1873. In the early morning hours of February 3, 1880, a tremendous fire burned the building to the ground. L. W. Welch; his elderly parents, Anna and Jacob Welch, and their grandchild; and Dr. Timothy Smith, his wife, and two small children, all barely escaped with their lives from their second floor living quarters. Two of the store's employees, Miss Tillie Irving and A. F. Hopkins, were trapped on the second floor and perished. This fire and another two years later at the Keystone Block led to active debates about the need for a municipal water system. Another step toward a safer town came in May that same year, when a "Hook and Ladder" company was organized and equipped. Watrous listed the officers as: "...John Place, Foreman; W. T. Seamans, First Assistant; E. M. Pelton, Second Assistant; W. P. Keays, Secretary; and John Deavers, Treasurer."[4] The group began training to be an effective fire-fighting force, but the lack of a system of pressurized water made their task more difficult.

Town Considers Water System Proposals

Town folks began circulating a petition in March 1880 to ask the town's Board of Trustees to consider building a gravity-flow water system that would take water from Claymore Lake, four and a half miles above the city. A *Fort Collins Weekly Express* editorial described the proposal:

> The latest enterprise started in this go-ahead town is a system of water works. Although there is plenty of water rushing through the town, but little of it is fit for domestic use, while in the case of fire it can only be utilized to extinguish flames by dipping it up with buckets, on account of the lack of both hydrants and steam engines....It is proposed that the town shall vote bonds to the amount of $60,000, to run ten or fifteen years, for the payment of the work....

The article also described the uses of this "water works" in detail:

> Water will be taken from Claymore Lake,...at an altitude of 150 feet [above Fort Collins], in iron pipes twelve inches in diameter. The water will be distributed, after reaching the city, in two-inch pipes, and hydrants will be established at convenient points along the streets, from which to obtain water in case of fire. Water will be carried to all the houses in similar pipes, to supply housekeepers. If the water from Claymore Lake, upon analysis, should be found unfit for household use, water would be taken from the Poudre at a point near Claymore Lake, a reservoir built and the water taken to the city as before described; but the elevation would be less, and consequently the pressure would be lighter....[5]

The proposed plan seemed to have all the elements for success since it would "....not only sup-

Above: Fort Collins Hose Team, 1887. The popular running team prepares for a race at the Larimer County Fair on September 16, 1887. Poudre Valley Hospital on Lemay Avenue now stands on the site of the old fairgrounds. Courtesy, Fort Collins Public Library.

Right: Fort Collins Hose Team, 1888. Team members look self-assured in this formal portrait. Courtesy, Fort Collins Public Library.

ply the town with plenty of good water, but also ensure the greater safety of life and property, and thereby induce capitalists to invest money here more readily. It would also tend to reduce the rates of insurance...."[6] So why did it take two and a half more years to get the citizens to approve a bond to build a municipal water system? Apparently, this original proposal never received much attention since there were no newspaper follow-up stories about it.

The question of a municipal water system resurfaced in late December 1880. In a letter to the *Weekly Express* editor, someone identified as "M" wrote about the need for "...a plentiful supply of wholesome water for domestic purposes...." "M" noted "...the prevalence of typhoid fever...." and the need "...for extinguishing fires.... It would have a powerful influence in bringing immigration and capital into our town and vicinity...." as reasons for building a water works. Other factors pointed out were the freshness of the Poudre's water, sufficient fall for water pressure, and losses from fires. "The loss sustained by fire within the last year in Fort Collins would go far towards defraying the expenses of efficient water works...."[7] The letter seemed to express the view of many townspeople.

The year 1881 brought other proposals, as "experts" came to Fort Collins to put forth their ideas. In late January, the John Mills Water Works Company made a pitch that promoted the use of the Mason and Hottel millrace as a power source for a turbine water wheel to pump water from the river. Other proposed features included two pumps, each capable of pumping two million gallons in twenty-four hours; two miles of pipe, with six-inch pipe along the main streets and four-inch pipe for the side streets; fifteen hydrants, ten double and five single ones; and one thousand feet of hose with a hose cart. The proposed cost was $80,000 financed with bonds at eight-percent interest for fifteen years.[8]

Two weeks later, the *Weekly Express* printed information about another "water works scheme." E. S. Alexander of Russell & Alexander from Colorado Springs, who eventually would get the contract to build the Fort Collins Water Works a year and a half later, made his firm's proposal. Since the town fathers had set a limit of $40,000 that they were willing to spend on a water system, Alexander presented a scaled-down version for the town. His company was well qualified for the job, having previously built systems for South Pueblo, Leadville, Colorado Springs, Silver Cliff, and Golden. Under his plan, water would be drawn from the town ditch (today's Arthur Ditch) at Sherwood Street and Mountain Avenue. The ditch, he suggested, would need to be deepened by two feet. Water would be taken through a large pipe to the rear of the new Welch Block, where a pump would be installed. It would be capable of supplying one million gallons of water every twenty-four hours with sufficient power to deliver an eighty- to ninety-foot stream of water from a hydrant. The system would have four miles of pipe varying in size from four to ten inches, and twenty hydrants. Firefighters would have fifteen hundred feet of hose and two "hose carriages." [9]

An election was held on May 11, 1881, on the water works question. The water works issue received 180 total votes but lost by eight. The *Weekly Express* editorialized:

> We do not think that the vote is a true indicator of the sentiment of the people of the town at large upon the question. In the first place, the matter had not been placed before the public in a manner to permit all to vote intelligently upon the subject, while many who were conversant with the matter thought they saw a job in it.... It seems to us that the matter should have been more thoroughly agitated. The people should have had some idea of the system to be adopted and the probable cost thereof.[10]

The water works was by no means a dead issue. During the next year, there were more proposals and calls for votes on the water works question. Some proposed building a mountain reservoir because of the problems of river water quality and seasonal volumes of flow. Another *Weekly Express* editorial in August described the problems:

> The frequent storms that have occurred up in the range recently have filled the river with large quantities of impure mat-

ter. First, the water runs red, there having been a flood in the red stone districts; then it is black, from the burnt districts; Thursday morning it was densely muddy. It was as bad in the river as in the ditches, and our water man was obliged to fill his cart [water wagon] from some well whose predominating ingredient was alkali. The river water was not even fit for cooking or washing....[11]

The People Finally Vote "Yes"

The new year, 1882, dawned with an increased awareness of the need for a town water works. The newspapers kept the issue in the public's eye with little editorial one- and two-liners. For example in January, the *Fort Collins Courier* carried in its "City News" column, "The question of an adequate water supply for Fort Collins is boldly forcing itself to the front."[12] Fires were still the main concern of the citizens with relief being expressed that there had been no major fires since the Welch building burned in 1880.

City Engineer H. P. Handy wrote a long, rambling two-part article for the March 2 *Weekly Express* about the various reasons that Fort Collins should have a water system. He pointed out the dangers of fires, problems of water-borne diseases, and needs of the town to become more modern. The town's Board of Trustees upped the ante by voting to allow $60,000 to be used for a water works. Beginning in March, the *Weekly Express* then began a series of articles about the need for a water system that led to an April vote on the issue. One article noted that the editor believed the people favored building a water works. The *Express* even sent a reporter out to interview business leaders around town. To a man these "leading citizens," including former state Senator William C. Stover; hotel owner B. S. Tedmon; businessmen David Patton and H. E. Tedmon; banker Noah Bristol; rancher Charles Ramer; and Mayor Jacob Welch, stated they would vote for a water works. Their comments were published March 30.

The vote was held April 4, 1882, and the results were published in both the *Weekly Express* and the

Fort Collins Fire House on Walnut Street, ca.1889. Volunteer firemen and their equipment pose in front of the1882 firehouse. City hall was on the second floor. Besides the hose team, the city used horse-drawn fire-fighting equipment. Courtesy, Fort Collins Public Library.

Courier. The results showed that the citizens did, in fact, favor supporting a water works. The vote totaled 268 in favor and 44 against financing a water system. It seemed that now the town could have its water works, but several weeks passed with no public action. Finally, on May 25, 1882, the *Weekly Express* printed the reason:

> The water works project has fallen through. The town board advertised for bids and the bids were opened. All but one bid was rejected. This was the Mills bid, and the plan was to construct a filter from the head of the Mason & Hottel mill race to the Mason & Hottel mill and force the water through town by means of a pump and engines, the former to be propelled by means of water power from Mason & Hottel's. We think we see in this the same old job that was attempted a year ago and very decidedly sat down upon....[13]

The article noted that "...the result is that there is the strong probability of the water works question falling through this year,"[14] which must have been extremely frustrating for citizens who thought their vote finally settled the issue.

By early June, a water works battle was shaping up. The *Weekly Express* published articles and letters to the editor opposing the Mills' plan that benefited Hottel, rather than the town as a whole. They wrote that the plan carried no guarantees that it would be adequate for providing both household water and water to fight fires. The *Courier's* editor, on the other hand, believed the town board was doing what the voters had instructed them to do—find a suitable water system for the town. The animosity between the two opposing editors was easy to discern. Ansel Watrous, editor of the *Courier,* made his feelings crystal clear:

> ....The *Express* has a right to oppose, but when that opposition is connected with low and vicious personalities it is subject to criticism....Where [editor of the *Express*] Crafts' ignorance and lying propensities are known, his articles do no harm; but to strangers and the outside

Fort Collins Hose Team No.1, ca. 1888-1898. The running team tradition continued even after adoption of horse-drawn equipment. The local men competed with similar Colorado fire-fighting teams, and this group received a ribbon for its July 4-5, 1898, performance. Courtesy, Fort Collins Public Library.

> world they are liable to create a wrong impression....[15]

Editor H. A. Crafts fired back with "Editorial Notes" in the *Express:*

> The people say "you are right," go ahead.
>
> We are in favor of water works—of the right kind.
>
> Public sentiment fully sustains the *Express* on the water works question.
>
> Nine out of every ten of the citizens of Fort Collins heartily support us in our position on the water works question.
>
> 'Hew to the line; let the chips fall where they will.' That will be our motto in dealing with this water works question....[16]

The newspaper rhetoric continued through June, with little about the efforts of the citizens or the Board of Trustees to resolve the issue. Nevertheless, all discussions became moot by mid-September.

Town Builds New City Hall and Firehouse

As the controversy and heated debate raged on, the city completed a new City Hall on Walnut Street in the summer of 1882. The narrow building housed fire wagons on the lower floor and city offices on the second floor. The elegant, red brick structure was crowned with an ornate bell tower. (Later, the bell tower was removed, but the Walnut Street building remained the seat of municipal government until 1957 and the home of Fire Station #1 until 1973. During building renovation in the early 1980s, private owners rebuilt the bell tower.) An *Express* article described the new fire bell:

> The new fire alarm, by the way a much needed public convenience, has been placed in position in the tower erected for its reception and is ready for use. The bell weighs some 1900 pounds and is of good bell metal, and will probably prove a sufficient alarm for our city for many years to come. The bell is stationary, with a swinging hammer, to which are attached two ropes, from either side,

Linden and Jefferson streets, 1884. Looking southwest, many imposing brick buildings lined the business district by 1884, but Frank Stover had not yet rebuilt his Keystone Block. During the September 1882 fire that destroyed two new buildings at the corner of Linden and Jefferson streets, firemen placed ladders against the single-story Poudre Valley Bank building at 233 Linden Street (bottom center) and carried hoses, fed by pumps at three nearby businesses, over the rooftops to gain access to the fire. The former bank building is now a two-story structure. An advertising banner is suspended over the wide street. Courtesy, Fort Collins Public Library.

> extending to the hallway on the lower floor, thus allowing the person who rings the bell to strike the alarms in rapid succession. This bell will prove a great convenience to the public generally, and to the 'fire laddies' particularly, who are delighted with this acquisition to their outfit.[17]

The "fire laddies," or firemen, must have been very popular with the town folks because the *Courier*, in its June 29 edition, decided to have a little fun with them:

> If there is one thing this gallant company excels in, it is practicing. They practiced on a team, in the street, and on the feelings of all who beheld them. No outside team wants any "truck" with them. The ladder was elevated in front of the Tedmon House and gallantly scaled. We suppose that is the right name for it, as from the reckless manner which they slid down it, it must have been a scaly performance. From the way they puff themselves up, we conclude that they are not in need of a newspaper puff and we will not give them any.[18]

The Hook and Ladder boys and a host of other volunteers were sorely tested in mid-September, and their inadequate defense against a major conflagration dramatically brought the need for a water works back to the public eye. And this only two days after the town trustees decided to hold another election September 20, on a "gravity plan." Frank Stover's nearly completed two-story Keystone Block, on the southwest corner of Linden and Jefferson streets, caught fire in early morning of September 15, 1882. In a short time, the fire spread to the adjoining building owned by T. H. Robertson and W. S. Haynes. A lengthy article in the next day's *Courier* described the fire and the efforts to fight it. The paper noted "....the hooks and ladders were promptly on hand...." and that "....in five minutes from the first alarm it was evident that, with the limited water facilities, [i.e., the bucket brigade] nothing short of a miracle could save the building...."[19] Here was a fire that threatened several

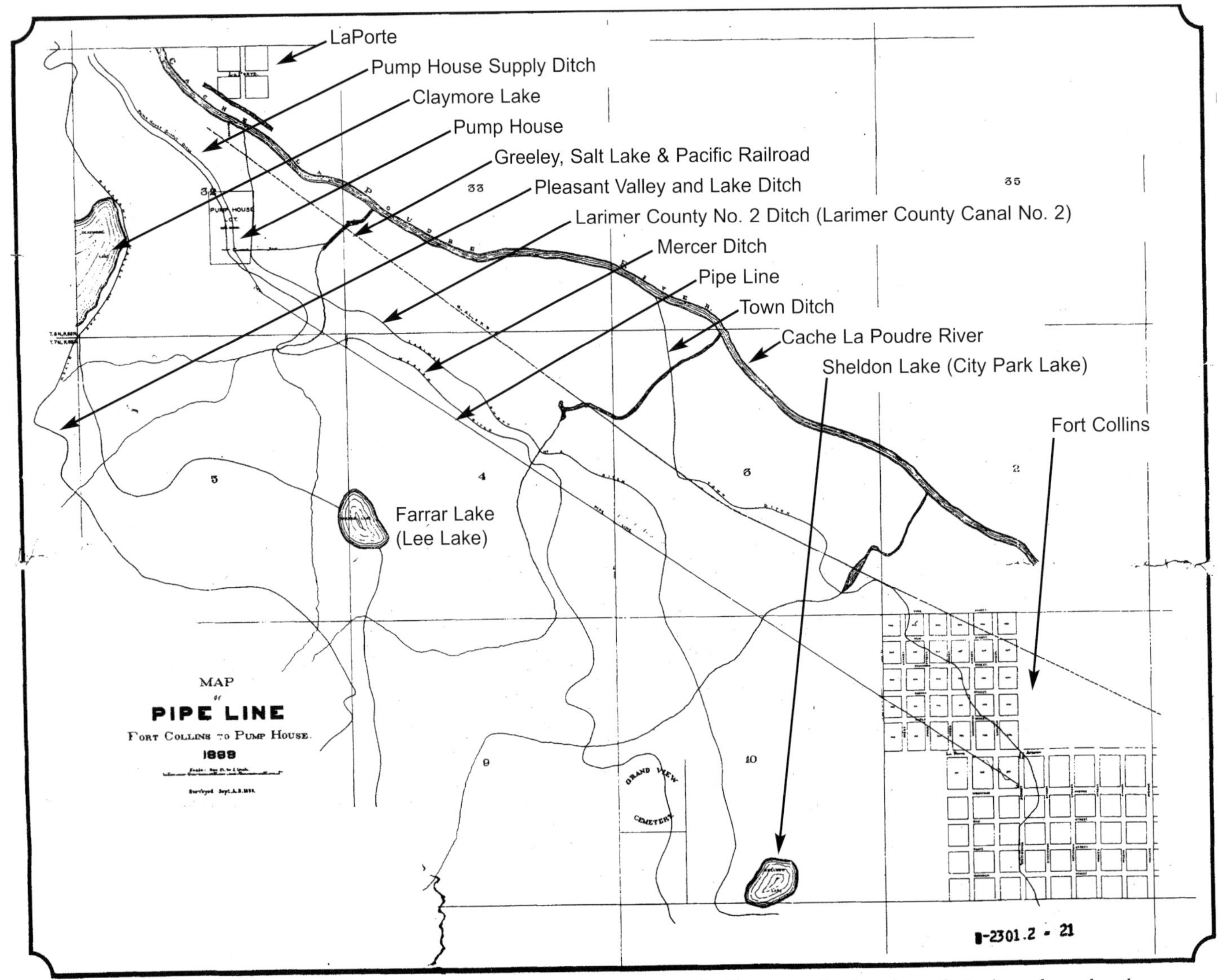

"Map of pipe line, Fort Collins to Pump House, 1889." The map not only illustrates the location of pipelines but also the irrigation canals that ran through Fort Collins. Courtesy, Fort Collins Water Utilities Department.

buildings in the small business district of one of Colorado's up-and-coming towns. The paper described what it called the "Herculean efforts" to save the buildings:

> ....Ladders were at once placed against the Poudre valley bank [sic][†], and its roof, and that of the Odd Fellows' hall, literally swarmed with men. Lines of buckets were formed from every available source of water supply, and a steady stream of water was soon pouring upon the roof of the Robertson building. Hose was attached to the force pumps of Rhodes and Tedmon on Jefferson street, of Abner Loomis on Linden street, and of Ames Patton & Co., on Walnut street.... Submitting to the inevitable, but not before it was inevitable, the attention of the gallant firemen was concentrated upon saving the Odd Fellows' hall and the Poudre valley bank [sic]. An unceas-

† At the time of the fire, Poudre Valley Bank was located at 233 Linden St., on the north side of the alley.

It was City Engineer Henry P. Handy who selected the site for the water works, laid out the route for the ditches, and was engineer in charge of construction. The equipment selected, construction materials used, and building techniques were all on the cutting edge of technology for the time.

—David Budge,
"Nineteenth-century Technology and the Water Works Site"
(Unpublished paper, Fort Collins Water
Works Archive, 2003).

> ing stream of water was poured upon their roofs by hearts that knew no faltering and hands that knew no fatigue, and their gallant and loyal efforts were successful, and the progress of flames were checked at the fire wall between the Odd Fellows' hall and the consumed Robertson building....[20]

Robertson's and Haynes', and Stover's buildings were completely destroyed by the fire. The lack of an adequate municipal water system made this kind of destruction inevitable. Another problem was that the water buckets proved to be hazardous as well. Dropped buckets injured several fire fighters. The same *Courier* article included a dramatic, but somewhat amusing story of a "lather," who was at work installing laths before plastering the walls on the second floor of Stover's Keystone Block when the fire broke out. The paper noted that he "....was so terrified by the leaping and roaring flames that he let go all holds and tumbled out of a window, landing on the edge of the stone sidewalk from whence he rolled into the open basement. Terror lent wings to his hands and feet and he quickly scrambled up to the street again and made off. He was slightly scorched and badly scared...."[21] One can only picture the poor worker, clothes smoking, dashing with a limp down Linden Street, perhaps headed for the Poudre River to cool off!

The *Courier* also stuck in some "I told you so!" statements that day:

> Instead of being a burden the water works will be a buoy.
>
> The interest of every man, woman, and child in Collins demands the speedy establishment of water works, and we think that interest will find a triumphant expression at the polls on the 20th of the month.[22]

New Vote Also Favors Water Works

The town's men (no women's suffrage yet!) turned out for the election on September 20 with 278 casting their votes; 182 favored the proposed water works at a cost of $85,000. The next day's *Express* put the vote into proper perspective:

> The majority rules. The majority says we must have water works.
>
> Better late than never. This is what the town board thinks about water works.[23]

A week later, most of the town's taxpayers met at the Opera House to discuss some of the different plans. The following day, the town trustees, including A. J. Ames, Jay H. Boughton, George S. Brown, Abner Loomis, Frank P. Stover, H. E. Tedmon, and William F. Watrous, accompanied by several engineers, made their way up the Poudre River to "....run some measurement lines on the proposed WW—and to better acquaint themselves with the lay of the land...."[24] A story in the October 4, 1882, *Weekly Express,* reported the results of this foray. One of the sites visited was Farrar Lake, three miles northwest of town and a half-mile south of the Poudre River. (Modern maps do not show this name because its was changed to Lee Lake many years ago.) The town fathers quickly discarded the idea of a gravity flow system from the lake because water had to be purchased from the Pleasant Valley ditch. Maintaining a winter flow in the ditch would create another problem. In addition, the fall from Farrar Lake to town was insufficient and would not provide enough water pressure.

The fallback plan was one that would build a two million gallon reservoir closer to the river, bring water to town through twelve- or fourteen-inch pipe, and build a "stationary engine" near town in case of fires. Bids went out immediately with an October 20 deadline for returning proposals. The 1882 town trustees had investigated and reported through the same *Weekly Express* article that:

> ....It is proposed to use wrought iron pipe, such as is made by the National Tube works, for the mains from the source of supply to the distribution point, and cast iron pipes for distribution purposes. The wrought iron pipe will be used as a matter of economy, as that, 12 inches in diameter, will cost but $2 per foot, while cast iron of the same size will cost $3.50 per foot.[25]

Engraved sandstone. Masons installed a commemorative stone tablet over the entrance to the pump house. Chiseled into native sandstone are the names of the people who oversaw construction of the 1882-1883 Fort Collins Water Works building and the water delivery system. The list includes Mayor George S. Brown and Town Trustees Frank P. Stover, Abner Loomis, William F. Watrous, Jay H. Boughton, H. E. Tedmon, and Andrew J. Ames; attorney Eph. Love; engineer H. P. Handy; contractors Russell and Alexander; and county treasurer Charles H. Sheldon. Courtesy, Fort Collins Public Library, Doris Greenacre collection.

Chapter 3: Building Fort Collins' First Water Works

The "Proposal for Bids" was published in the two local newspapers and sent out to prospective bidders in early October. The request included a completion date of, "....on or before June 1, 1883; but 1,000 feet of street mains, ten double hydrants, engines and pumps to be placed and in working order within 60 days from date of contract...."[26] The town could do nothing more until the October 20 submittal date but hope that no big fires occurred. The town board met shortly thereafter and heard a report by an engineer hired to draw up an estimate of the cost for the proposed water works. The engineer did not think the gravity system would be sufficient for fire protection and proposed a combined gravity and direct pressure system. City Engineer Handy and the trustees accepted this advice.

On October 30, the board opened four competing bids. Russell and Alexander; South Pueblo Machine Company; Davis, Creswell and Company; and H. H. Given and Company submitted proposals. The trustees decided to tweak the plan a little after going through the bids and ask for some changes. The revised plan would, "....run pipes up the river to the neighborhood of Laporte, which would give purer water and 75 feet of head [pressure], instead of 55; also to station a pump to be propelled by a turbine wheel about two and one-half miles above town for additional pressure."[27] The bidders must have quickly reworked their bids because two days later the town board chose Russell and Alexander as the contractors to build the Fort Collins' system at a cost of $77,000. And, they promised to complete the work by April 1, 1883.

At long last, work could begin! Bids also were solicited for, "....earth and gravel excavation on reservoir and race [ditch],..."[28] By November 13, the *Express* reported:

> The contract for excavating the reservoir and water-power ditch for the Fort Collins water works was let to L. A. Clark, and ground was broken today. The contract calls for 30,000 yards of earth work. Mr. Clark is working ten teams, and will have the reservoir done this week if the weather is good. The survey is being prosecuted as fast as possible. City Engineer Handy has a field party at work, and will complete the survey this week, with the exception of the pipe lines. The contractor will probably go to work on the water-power ditch next week.[29]

A committee of town trustees, composed of Loomis, Boughton, and Watrous, looked over the route for the right-of-way the following week. They recommended some changes that would save a little money. The location of the new "pumping work" was moved to a site two miles upstream, near where the reservoir was being built. It was situated on "the Chamberlain place," while the settlement pond was on a farm owned by the Greeley, Salt Lake, and Pacific railroad, a subsidiary of the Union Pacific railway.[30] The railroad's main line, completed in April 1882, ran on the south side of the river and went into the Horsetooth valley where the Stout and Redstone quarries were located. This change finally placed the water works, reservoir, and water-power ditch on the present-day road called Overland Trail.

During the next few months, the *Express* and the *Courier* printed frequent progress reports for the townsfolk, some of whom were eagerly awaiting the day they could tap into the new water system and bring fresh water into their homes and businesses.

> November 25, 1882—City Engineer Handy informs us that 1900 feet of the water-power ditch have been excavated,... The settling reservoir has been completed.[31]

(This reservoir was located about a mile above the pump house.)

> December 5, 1882—Russell & Alexander, contractors for the water works, were empowered to extend the street mains up Howes and Matthews [sic] streets, which will give an additional number of hydrants and no "dead ends," but a free circulation through the town."[32]
>
> December 6, 1882—Mr. E. S. Alexander, of Russell & Alexander...has opened an office in the Jefferson street block.... Excavation for the pipe lines will begin in about ten days....Thirteen-hundred feet of pipe, about one-forth [sic] of that required, has arrived....[33]
>
> December 18, 1882, headlines—Laying of Street Mains Commenced on College Avenue To-day. A Large Force of Men at Work Under Superintendent [J. E.] Ford.[34]
>
> January 23, 1883—Eight car loads of stone for the foundation of the Water Works' Pump House were taken from the Stout's Quarry to Laporte on this morning's train.[35]
>
> January 26, 1883—J. J. Bradley is at work to-day upon a stone tablet, 6x4 feet, which is to be inserted in the front wall of the water works pump house. The name of the mayor, city council, city engineer, and the contractors for putting in the water works are to be chiseled upon its face.[36]
>
> March 22, 1883—Water service is being attached to [A. H.] Patterson's building in the Jefferson street block.[37]

(Patterson's may have been the first business to be "hooked-up" to the new water system, sans water for a couple more months!)

The promised completion date of April 1 was approaching rapidly, and the machinery for the pump house had not yet arrived. The *Express* spoke to the frustration of the entire community when it

The foundation of the original Water Works reflected the heavy-duty usage expected from the building. The sandstone foundation and the piers for the machinery had to withstand the movement of water and machinery. People in the building industry were convinced that sandstone from the Fort Collins quarries produced good foundation stone. Frederick Spalding in Masonry Structures *wrote that the better grades of sandstone reach a compressive strength of 6,700 to 19,000 pounds per inch, and a modulus of rupture of 500 to 2,200 pounds per square inch in transverse strength. This compared well to granite or marble, but was lighter in weight, locally available, and sufficiently strong for the Water Works application.*

About the time the pump house was under construction, the Union Pacific railroad began to develop its quarries in the Fort Collins area to supply light-colored and pink sandstones for its own railroad construction and for impatient buyers in Kansas City, Omaha, Topeka, St. Joseph, and other towns. So the sandstone foundation of the pump house tells a story of more than just local usage. It is a connection that interprets the intertwined growth of Colorado railroads and industry in the late 1870s and 1880s.

—Susan Quinnell,
"Historical Interpretation Report of the Building Materials of the Fort Collins Water Works Building, 2005 North Overland Trail"
(Unpublished paper, Fort Collins Water Works Archive, 2001).

published the following comments in its March 31 edition: "Over $60,000 worth of property has been destroyed by fire in this city since the water works were voted,...It is confidently believed that seventy five per cent[um] of the losses occasioned by this morning's fire could have been prevented had the waterworks been in operation."[38] The machinery arrived on April 9, consisting of,

> ....four large Gaskell [sic] pumps, manufactured by the Holly company at Lockport, N.Y., and two large-sized American turbine water wheels, and an improved regulator, which automatically adjusts the supply with the demand. The machinery is of a heavy class and appears to be well and substantially made. The capacity of the pumps is one and one-half million gallons per twenty-four hours....The arch and foun-

Stone foundation and brick exterior wall. "Bushed" stonework, a common decorative treatment with ancient origins, demonstrates the builders' concern for aesthetics as well as for utility. Laborers first used stone saws to make a smooth, planed surface and rubbed away any roughness with grit. Stoneworkers then created the stippled surface by using a "bush hammer," a hammer with a diamond-pointed face like a meat tenderizer. They also used a stone chisel to make the diagonal scoring marks. Fort Collins Water Works Archive, Susan Quinnell photograph.

Scored mortar. Striking the mortar to produce a horizontal line, as seen in the pump house exterior, was a common stylish embellishment of the nineteenth century. Bricklayers used the American common bond method to build the pump house walls. In this method, the bonding course of headers (using the short end of the brick), appears once in every five, six, seven, or even eight courses. The great variation in the placement of header courses creates an informal appearance, typical of late-nineteenth century picturesque architectural revivals. Fort Collins Water Works Archive, Susan Quinnell photograph.

Above: Windowsill detail. Windowsills also received bush-hammer treatment. Fort Collins Water Works Archive, Susan Quinnell photograph.

Left: Tudor-style window crown. The oversized, double-hung windows of the pump house are especially picturesque and bold, and fit well into the structure's rural setting. Single window units were placed symmetrically apart, common in vernacular architecture, and allowed even lighting in the work areas. Modest ornamentation around the windows, including the Tudor-style, segmented sandstone window crown arches, offered the bold delineation favored in Gothic Revival and other picturesque styles. Fort Collins Water Works Archive, Susan Quinnell photograph.

> dations for the pump house will be completed tomorrow, and it is probable the pumps will be placed in position immediately,...[39]

The reports during the next two months could now be more optimistic as more pipe and building supplies arrived, and the progress at the pump house was more visible. Pipe laying from town to the pump house had been completed, and the laying of street mains from the West Mountain Avenue main connecting point was progressing well. Sale of the municipal water bonds seemed stalled, until the contractors, Russell and Alexander, agreed to take them at a discount. "The price received was about 91½ cents on the dollar of the $85,000 worth of bonds issued. The price will net the city about 90 cents, the remaining 1½ cents having been used in the way of 'commissions,' etc.," reported the *Express* on April 16. Russell and Alexander must have had a better financial reputation than the small town of Fort Collins, for the paper continued, "We under-

Stone pillars. A Colorado State University anthropology graduate student points out an embedded bottle between the stone piers. Archaeologists exposed four sandstone piers used to house the original pumps during the 2001-2003 excavations inside the pump house. Two piers were disturbed and two were left in good condition when the pumps were removed in 1920. The disturbed piers show evidence of large bent rods used to secure the pumps. Courtesy, David Budge.

Cast iron flanged pipe. Archaeological crews discovered this ten-inch diameter cast iron pipe in the southwest corner of the pump house during excavations in 2000. Wrought iron pipe is attached to the cast iron flange west of the building. Wrought iron was lighter than cast iron and just as strong. Notice the tree root extending through the pipe. Courtesy, David Budge.

Fully excavated cast iron pipe. Crews discovered a five-inch diameter tree root that reached through the pipe and downward in search of water beneath the building. In order to proceed with pump house excavations and wall stabilization work, the crew removed the root in 2003. Courtesy, David Budge.

stand that the above purchasers have already disposed of a large proportion of the entire amount."[40] Another unexpected expense came up the next day:

> Mayor [Abraham L.] Emigh called the attention of the council to the necessity of making suitable provisions for carrying the water of the tail race of the pump house under [Larimer] canal No. 2, and said that the ditch owners insisted upon this being done immediately, so that they could turn the water into their ditch.[41]

City Engineer Handy addressed the problem and stated, "....the difficulty complained of by the ditch owners could be readily adjusted by the expenditure of about $65."[42]

The Larimer County Canal No. 2, at this time, ran parallel to the water works' supply ditch, about two hundred yards east and intersected the discharge "tail race" perpendicularly before meandering east toward town. It was not until 1906, after the Poudre Canyon Water Works was built and the old plant was no longer in active use, that the City of Fort Collins leased the supply canal to the Larimer County Canal No. 2 Company. The original ditch was filled in and partly covered by the present Overland Trail.

Ansel Watrous, editor of the *Daily Courier*, accepted the contractors' invitation to tour the system and to see construction progress at the pump house. He wrote a long, rambling article in the April 19 *Courier*. His detailed description gave the first complete look at the work that was being rushed to completion:

> PUMP HOUSE....a brick structure resting on a stone foundation of solid masonry, 22x28 feet square and 14 feet high from floor to ceiling. In this building, the four Holly pumps will be placed on stone piers, the foundations of which extend below the foundation of the building. These piers, four in number, are each twenty feet high and three by nine feet in size, and extend some six inches above the floor of the pump house. The cap stones on the piers are in one block and from eight to ten inches in thickness.... Underneath the pump

American Water Wheel Company turbines manufactured by Stout, Mills and Temple were installed at the 1882-1883 Fort Collins Water Works. So far, attempts to locate design and operating specification criteria on this brand of turbine have been unsuccessful. However, James Leffel's illustrated handbook, Improved Double Turbine Water Wheel for 1881 and 1882, *makes it possible to estimate design and operating standards because the various manufacturers used common patterns in the 1880s.*

Newspapers in 1883 state a capacity of 2.8-million gallons per day at the pumps and discharge in Fort Collins of 1.3-million gallons, or a flow of about 903 gallons per minute. The waterpower requirements to produce this flow can be estimated from Leffel's book. A 75-horsepower American turbine operating at maximum with a design head of 22 feet, such as was used at the Fort Collins Water Works, is similar to the Leffel 35-inch size that produced 79.2-horsepower and turned at 185 rpm, requiring 2,147 cubic feet of water per minute.

Newspapers indicate that the American was frequently set to operate at 15 horsepower. The Leffel 35 generated 14.2 horsepower at 104 rpm and required 1,211 cubic feet of water per minute. The lack of suitability for this type of technology in a semi-arid desert environment, such as the Cache la Poudre River basin, can be illustrated by the following considerations.

A cubic foot of water weighs 62.5 pounds and represents a volume of 7.5 gallons. This means that between 9,082.5 and 16,102.5 gallons per minute or 544,950 and 966,150 gallons each hour were required to produce 15 or 75 horsepower per wheel, respectively. In a 24-hour period each water motor would have required either 13,078,880 or 23,187,600 gallons of water, depending on the running horsepower. Operating together the low and high horsepower water requirements would have been 26,157,760 and 46,375,200 gallons daily.

To summarize, to send one gallon of water to Fort Collins the turbines required about 31 gallons! And, the city needed 1.3- or 1.5-million gallons per day. What worked well in the humid Midwest and East Coast regions was at best marginal or woefully inadequate in dry Colorado.

Actually, the plant may have functioned quite well, however, and not required a lot of turbine power due to the 85-foot difference in elevation between the pump house and Fort Collins. For each 2.31-foot difference in elevation, one pound of water pressure is generated by gravity. Accordingly, almost 37 pounds of water pressure would be generated in downtown Fort Collins.

—David Budge,
"Nineteenth-century Technology and the Water Works Site"
(Unpublished paper, Fort Collins Water
Works Archive, 2003).

house is a subterranean archway twenty-two feet in length with fourteen foot breast and seven foot spring, of immense strength, built of stone, each stone being cut and fitted for its own peculiar position and each forming a key stone for the whole structure. The side walls of the arch are three feet thick and the foundation walls four and a half feet thick, deeply imbedded in the ground. Under this archway the water that passes [through] the water wheels escapes and finds its way back into the river through a long tail race. All the work above mentioned is now well advanced toward completion....

SUPPLY CANAL AND RESERVOIRS. The supply canal taps the river about 100 feet below the head gates of the Mercer ditch, and passes under the Greeley, Salt Lake, and Pacific railroad a few rods east of the Point of Rocks....

Pump room area. Crews removed many cart loads of fill dirt and stone to reach this level. Courtesy, Richard Carrillo, Cuartelejo HP Associates.

Pump room features. A mason evaluates the brickwork in the pump room in December 2003 before beginning repairs. During the 2001-2003 excavations, archaeologists, field technicians, and volunteers excavated the pump room so that structural repairs could be made. This work revealed the top of a stone archway on the west wall (bottom left), a large water pipe, like the one in the southeast corner, that extends into the room from the west (center), and one of the stone piers that held a pump (foreground). Courtesy, David Budge.

The head gates flume is fourteen feet in width....

The water, immediately after passing under the railway, is discharged into a large settling basin. Thence it is conducted along the base of the bluff, and just below the Mercer ditch, at a grade of nine inches to the mile, a distance something over a mile to the crossing of the Laporte road over the Mercer ditch, into the main reservoir. This reservoir is 220 feet in diameter, and eight feet deep. The

Diversion structure. The stone retaining wall is part of the original Fort Collins water delivery system located on the south side of the Cache la Poudre River above Laporte on private property. Builders of Fort Collins' first water system captured water from the river by using the same method as early agricultural irrigators: they built a dam to divert water into the supply canal that transported water to the pump house. Courtesy, Wayne Sundberg.

capacity of the reservoir, canal and settling basin is, in round numbers, 4,500,000 gallons, a supply sufficient for any emergency....

FILTER. In the center of the reservoir will be constructed a filter, through which all the water supplied to the city will pass. The structure will be 16x36 feet on the outside, constructed with double walls made of 2x8 plank, laid flatways one above the other, with two inch interstices between each laying. A three foot, eight-inch space between the walls will be filled with gravel and sand, through which the water will percolate into a basin 4x25 and eight feet deep. From this the water is drawn into the supply pipes, and forced into the mains and distributed all over the city. Every precaution is being taken to make the city supply pure and fresh....

FLUME AND PENSTOCK. Leading from the reservoir to the penstock and forebay is a strongly built and substantial flume made of 12 x 12 timbers, framed and bolted together and planked with three-inch plank. The water to drive the wheels, which in turn drive the pumping works, is discharged from the flume into a penstock 10 by 16 feet square and 24 feet high. This is built of 12 by 12 timbers, substantially framed and bolted together, and planked up with heavy planking. The floors of the penstock are double, the lining being of three-inch plank, jointed beveling and fitted tightly, and then caulked with oakum, pitch and tar. The upper course is laid with two-inch plank, jointed, fitted and caulked in the same manner. The penstock and fore-bay rests on solid masonry, laid in English Portland cement. The forebay, where the water wheels will be placed, is 8 by 16, and is simply an extension of the penstock, but not so high, and constructed in the same manner. The wheels will

Railroad bridge over supply canal. Blocks of sandstone support the Greeley, Salt Lake, and Pacific railroad bridge over the supply canal. Now located on private property, the abandoned bridge was built in conjunction with Fort Collins' first water delivery system. The now-abandoned railway once delivered carloads of sandstone from the quarries south of Bellvue to construction projects in major Midwestern cities. Courtesy, Wayne Sundberg.

> work under a head of twenty-two feet, which will give them immense power, of which, however, only 16-horse power will be required for use under ordinary circumstances....[43]

This descriptive article must have given hope to citizens that they soon would have water flowing through the city mains to provide fire protection for the town and water for homes, should homeowners choose to pay for hook-up to the new system. The city trustees published a long list of annual water rates in the March 29, 1883, *Evening Courier.*

The optimism of the Watrous report may have been a little premature, for completion of the whole system was still several weeks away. Problems with the right-of-way across the land owned by the Greeley, Salt Lake, and Pacific railroad had not been resolved, so the supply canal could not be completed. Construction of the pump house was still unfinished. A brief April 25 article pointed out these issues.

> J. G. Lunn and men went out to commence the brick work of the pump house of the water works this morning. A half-dozen carpenters are now employed and everything is being pushed and a speedy completion of the works is looked for. Mr. Russell, of Russell & Alexander, informs us that their contract will be completed in less than two weeks. The railroad company has agreed to put in a bridge across the ditch above the pump house in accordance with Mr. Handy's plans. Of course the city will bear the expense of building this.[44]

The next day, the paper reported, "Engineer Handy says if there is no hitch a test of the water works can be made in two weeks."[45] The project was nearly a month past the original completion date proposed by the contractors the previous fall. Now, the "two weeks" projected by Handy and the contractors were still overly optimistic. There was still much to be done.

Supply canal. The Fort Collins Water Works supply canal meandered nearly a mile to reach the reservoir behind the pump house. This segment of the canal is now in private ownership. Courtesy, Wayne Sundberg.

Water System Tests

It was not until May 24 that the press could report, "the water works have been completed...."[46] The supply ditch was to be filled from the Mercer Ditch, but would not provide sufficient water to run both pumps. The city had to post a $1,000 bond with the court, in order to grant the city legal right-of-way across the railroad's land. Finishing the supply canal could now proceed. Finally, on the last day of May 1883, a *Daily Express* headline stated: "TESTING THE WATER WORKS. Preliminary Test Made Yesterday With Highly Gratifying Results."[47] At last, the townspeople could see the results of their long-awaited municipal water system.

> Messrs. Russell & Alexander, with the assistance of the hose company, made a very satisfactory test of the water works yesterday. The final test, of course, could not be made, in view of the fact that only sufficient water was running in the water power race to carry one set of pumps, and there was but sixteen feet of head, while the full head is twenty-two feet. Another thing was to be taken into account; the fact that all the machinery is new, the pipes rough, causing friction, and also leaking in many places. A pressure of 140 pounds was obtained to start with, and the hydrant at the Grout corner[‡] and one at the Loomis & Andrews block[§] were opened, with hose and nozzles attached. Two large streams were thrown simultaneously to a height of about 90 feet. The nozzles were pronounced by Mr. Russell to be too large for a good display, they being one and one-eighth of an inch in diameter, and "sprayed" the stream too much. The experiment was witnessed by a large number of people, and all expressed

‡ The Grout corner referred to the corner of Linden and Jefferson streets although the "Old Grout" building had been demolished to make way for Stover's store that burned in September 1882.

§ The Loomis and Andrews block was the newly completed three-story, brick structure at the corner of Linden and Walnut streets.

themselves highly pleased with the results....[48]

Only four leaks were found and repaired. A headgate was installed between the two ditches so water could be diverted from the Mercer Ditch, since the pump house supply ditch still was not finished from the settling reservoir back to the diversion structure on the Poudre River. A sixteen-foot long flume from the Mercer Ditch to the supply canal carried the water the short distance between the parallel canals. Still using only one pump, two hundred pounds of pressure could be produced, resulting in 120 to 130 pounds of constant pressure in town. On June 1, the *Express* reported, "....Mr. Handy says that the pipes will become clean so that the water will run clear within ten days...."[49] Full water service was coming closer.

Two more tests were made on June 4, one at the college, "....threw a stream one hundred feet in height for two hours...."[50] ("The college" refers to the State Agricultural College of Colorado, now Colorado State University.) Another, from a hydrant at College and Mountain avenues produced a column of water 115 feet into the air.[51] Two days later, a special edition of the *Fort Collins Daily Evening Courier* crowed, complete with a picture of a rooster gracing the front page, "The Official Tests of the New Fort Collins Water Works with Good Results" followed by "Six One-and-an-Eighth Inch Streams Thrown Nearly 100 Feet High."[52] The story covered nearly the entire front page and included a detailed description of the water works, "booming" the many attributes of the town of nearly fourteen hundred souls now far ahead of its neighbors.

The tests were conducted by attaching fifty feet of hose to hydrants at six locations using different size nozzles, and turning all six on simultaneously. The locations, at the corners of Jefferson and Pine streets, Jefferson and Linden streets, Linden and Walnut streets, Linden Street and Mountain Avenue, College and Mountain avenues, and "....in front of N. Weaver & Co.'s store on College avenue...," covered nearly the whole business district, and produced streams varying from 105 feet to 86.8 feet.[53] Had the whole business district been burning the "Hook and Ladder Laddies" would have been able to make a valiant stand against the flames. Fort Collins' residents now could feel free of such a conflagration, and since brick and stone structures were beginning to replace the original wooden ones, the city was now relatively safe from fires.

In addition, the city now had two fire companies, composed of thirteen and eighteen "fire laddies," respectively. Each company was supplied with a hose cart, one thousand feet of "linen hose," nozzles, ladders, and the rest of the equipment needed by an effective fire-fighting force. These men were also the fastest runners in town, since they were required to run to their hose company's location upon hearing the fire bell, pull the hose cart to the hydrant nearest the fire, hook up the hose, and begin putting water on the blaze. The rest of the company hitched horses to the "hook and ladder" wagon and brought the remainder of the fire-fighting gear to the site. They even competed with hose cart teams from other towns to establish which town had the fastest and most superior fire fighters. One "Hook and Ladder" company would have been stationed at the new firehouse on Walnut Street. The other was probably located somewhere west of College Avenue in the residential area. At this early date, only ten years after the town's 1873 platting and incorporation, the west boundary was Whitcomb Street; Laurel Street on the south, with only two buildings on the grounds of the Agricultural College; Cowan Street on the east; and the Cache la Poudre River on the north. With the variety of fire hydrant locations, two fire companies should have been adequate to protect the small but growing town.

On June 7, Russell and Alexander, ".... gave a farewell supper to the members of the city council, city officers, ex-members of the previous town boards, and invited guests, at the Tedmon house...."[54] Several members of the local and Denver press were in attendance. Praise was heaped on many people, including the two hosts. City Engineer Handy came in for much praise for originating many of the ideas for the system. Judge E. A. Ballard gave what was undoubtedly the most entertaining speech. He ".... commenced by witty criticism of the old board, saying that in their efforts to

Lawn sprinkling college dormitory, ca. 1880s. New lawns and trees surround the "college dormitory," later known as "Old Main," on the State Agricultural College of Colorado campus, now Colorado State University. Fort Collins' first water delivery system enabled the college to "sprinkle" new landscaping. Water pressure seems more than adequate to cover the grounds on the right side of the building. Fire demolished "Old Main" in 1970. Courtesy, Colorado State University Photographic Services.

benefit the city they had spoiled all anticipation of going to fires as they would be put out before one could get there. He thought the present city council equally open to censure for having made water too cheap and whiskey too dear...."[55] He also praised the contractors for their wonderful job.

The newly-elected town board met the next evening on June 8 to officially accept the water works from Russell and Alexander. Other items relating to the water works also came before the board, one being the appointment of J. W. Horn as the first pump-house superintendent, with pay of $75 per month. "....The contractors kindly consented to allow one of their men to remain at the pump house one or two weeks with the newly appointed superintendent...."[56] This way he could learn the intricacies of operating the new system firsthand. The trustees also discussed putting in a telephone or telegraph to the pump house. A fire alarm to alert Superintendent Horn if a fire was discovered in town was already on order. This warning device arrived the next day, and installation began immediately.

Citizens Slow to Sign Up

The first water taken from the new system by a city user showed a need that had not had much discussion when the water works was being planned and built. At the college grounds, "....A hundred foot hose was attached to a lawn sprinkler, and the lawn is receiving a good soaking. The force, gravity alone, is said to be very strong, sending the water a long distance from the sprinkler...."[57] The lawn was near the building newspapers referred to as the "college dormitory." The building also housed classrooms, president's office, and rooms for other uses. Eventually, it gained the name, "Old Main." Home and commercial users were slower to sign on than the college.

On June 12, the *Daily Express* reported, "Seventy-five applications have already been made for water service, which includes very few of the business houses, ..."[58] The city was still awaiting the arrival of taps and tapping machinery before the connections could be made. It was anticipated that application for about 130 taps would be submitted

Application for Water.

*No.*__________

*FORT COLLINS, Colo.,*__________*189*

The undersigned hereby makes application for the use of Water from the **Fort Collins Water Works,** *for the following purposes, on Lot*________*Block*________*:*

Kind of Building -------------	*Extras* --------------------
No. of Rooms ----------------	--------------------------------
Water Closets----- *Urinals*-------	BUILDING PURPOSES:
Bath Rooms ------------------	*Concrete,*------------*cubic yards*
Horses--------- *Cows*---------	*Stone Work,*--------------*perch*
Lawns -----------------------	*Plastering,* ------------*sq. yards*
Gardens,----*squares (of 100 ft. each)*	*Brick Work,* -------------*per M.*

I hereby certify that the above is a true description of the premises to be supplied, and the water will be used for no other purpose.

Application for Water, ca. 1890s. Home and business owners used this form to apply for a connection to the municipal water system. They were granted a water license that stated how the water would be used. Rates were set accordingly.

by the end of the year. During the next two months, the newspapers boosted hooking up with the new system. By mid-August, a rather scathing editorial in the *Daily Express* chastised those who had not yet applied for a tap.

> Over sixty applicants have now been supplied with water from the mains, and the city plumbers have about caught up with their orders. Still there are hundreds of houses on the line of the water mains which have not been connected with the works. Many who voted for water works do not avail themselves of the benefits now that they have been put in. They still take water from the cart, the ditches or wells impregnated with alkali. This is considered by many as economy. They would sooner suffer the inconvenience of the bucket supply, endanger their health and cheat the city treasury, than to spend a few dollars in having water pipes laid into their houses, which in one year will pay for themselves twice over, in convenience if not actual saving of money....No one once having water in his house will ever dispense with it. So we would urge

> upon our citizens, not only as a measure of health, economy and convenience, but in the spirit of patriotism and local pride to patronize the water works. They are not owned by a monopoly, but by the city, and the city first or last has got to pay for them. The man who takes his water from them not only secures a great blessing but pays a part of his taxes.[59]

Not all the problems with the new water works were as dramatic. Some were downright humorous. One newspaper reported that some were puzzled with the cloudy, bubbly appearance of the water one particular day.

> Some thought that a natural soda fountain had been tapped, while others suggested that a cow had accidentally fallen into the river above the head gate. Superintendent Handy, however, gave to a reporter the scientific and only reliable explanation of the phenomenon. 'Yesterday,' he said, 'the pumps were stopped and the supply for the city was drawn from the pipes alone by gravity. More or less air consequently got into the pipes. When the pumps were started this morning this air was forced into the water after the fashion of charging a soda fountain.' This was what gave the water the peculiar milky color. By letting the water stand a minute the air escapes and the water became clear as crystal.[60]

The system ran smoothly for the rest of the year. The supply canal was finished by mid-July, allowing both pumps to be used, if necessary. A home for the pump house superintendent was being built at the water works, at a cost of $1,000. In late January 1884, water rates were reduced, perhaps as an inducement for more people to attach their homes or businesses to the system. It must have worked. A year later, the *Weekly Express* reported these figures: February 1, 1884—132 users of the water works; February 1, 1885—230 users.[61] The system was building up to its potential.

Sprinkling unpaved street, ca. 1885. In order to control the dust from Jefferson Street, a worker sprays water in front of the Tedmon House. The luxury hotel stood at the corner of Jefferson and Linden streets. It was demolished to make way for the Union Pacific railroad's new line that paralleled Jefferson Street in 1910. Courtesy, Fort Collins Public Library.

Chapter 4: The First Ten Years of Water Service

Between 1883 and 1893, the new Fort Collins water system operated with minimal problems. There were a few leaks here and there, and a major break when a joint separated and caused a shut down that lasted several hours. Other than these types of problems, the system worked well. There were occasions when high waters in spring or heavy cloudbursts caused the water in the system to become discolored. On these occasions, the amount of mud flowing through the supply canal was more than the sand and gravel filter at the bottom of the reservoir could handle. When this happened, there were often calls for a better filtering system. Since the town had grown rapidly during the 1880s, more water mains were added to the new-growth areas south and east of the original town. In late summer 1887, a telephone line was extended from City Hall on Walnut Street to the pump house.

On the final day of 1887, the *Weekly Express* sang the praises of the system:

> Fort Collins has one of the best systems of water works in the west and the quality of water supplied by it to all parts of the city is unexcelled. It comes cool and clear from the mountain gorges. There has been hardly a fatal case of typhoid fever in the city during the five years that the system has been in operation.[62]

Further praise came from the *Weekly Courier* two years later, when the newspaper published a long article entitled, "Municipal Improvements." Under the subheading, "Water Works," the paper published a proud summary of the system at that time.

> Fort Collins built in 1883 and now owns and controls one of the best and most complete systems of water works to be found in the state, which supplies the citizens with an abundance of pure snow distilled water for domestic uses and for fire protection at a nominal cost. The daily average pressure on the pipes is about sixty pounds to the square inch. This can be increased in time of fire to 200 pounds, enabling the firemen to throw several streams to a height of 150 feet. The water is taken from the Cache la Poudre river at a point five miles above the city and conducted in pipes to the distributing mains. The entire cost of the system, exclusive of the yearly expense of maintaining the same and superintending the distribution of water, foots up at the present time to $165,000. The following is a statement showing the number of feet of mains of the different dimensions belonging to the system:
>
> | | | | |
> |---|---|---|---|
> | 10 inch mains | .... | 25,920 | feet |
> | 8 " " | .... | 5,166 | " |
> | 5 " " | .... | 8,660 | " |
> | 4 " " | .... | 3,452 | " |
> | 3 " " | ...... | 17,734 | " |
> | Total | .... | 60,941 | |
>
> To these are attached 45 fire hydrants, 23 water gates, 411 taps, and 462 stop boxes. The collections of water rents are increasing year by year and for 1889 amounted to nearly $9,000.[63]

Health Problems and the Town's Solutions

In late May 1891, citizens met with city council members to discuss a diphtheria outbreak in the city. While only four cases had been reported officially, as per a city ordinance, N. C. Alford, a former alderman, noted that many people were using "quack remedies" from "unqualified doctors," who "made no report to the city clerk."

E. W. Reed pointed out the unsanitary conditions on the streets and in the alleys of town. C. Golding Dwyre spoke about "....the dangerous condition of the privies and cesspools, especially in the business part of town." And, Dr. E. A. Lee, a member of the city's Board of Health, noted that "....the impure condition of the city's water supply had very much to do with the ill-health in the town."

The year before, $1,200 had been spent on repairs and cleaning the water system. Discussion focused on refinancing the bonds at a lower rate (five percent), to allow borrowing $30,000 for improvements at the water works. An expert water engineer brought from Denver to inspect the water works recommended changes that would, "....institute a complete system by which the water could be thoroughly filtered and delivered pure and clear in town [and] cost at least $20,000." Remarks by Dr. Lee, ... "who thought that the present filter used at the water works was worse than nothing and should be taken out and destroyed, ..." concluded the meeting. Those who attended the unofficial council meeting unanimously passed a citizen-initiated resolution to impose quarantines on homes where the disease was present.[64]

On June 18, the *Weekly Courier,* added its voice:

> The time is near at hand when the people of Fort Collins must face an unpleasant alternative. They must either provide means for improving the quality of water furnished to consumers, enlarge the cemetery to make room for the dead, or pack up and leave their homes for owls and bats to occupy.... The water people are compelled to use about one-third of the time is simply execrable [very bad] and is unquestionably the fruitful source of much of the disease and death from which this community has suffered the past year.... For a period of several days after a storm, washout, cloud-burst or flood the water is unfit for any domestic use and it should be labeled with a skull and cross bones.... The fault is not so much with the character of the water for

> purity, except during flood time, as with the provision for filtering it. The arrangement in present use is a libel on the name of filter, and should be condemned as a nuisance.... Let pipes be extended from the pump house along the bed of the present supply ditch to the river, and there connected with a system of underground chambers constructed in the bed of the stream, into which the water is permitted to filtrate through sand and gravel. In this way we can make sure of a constant supply of clear, pure water. It will cost money to do this, but it must be done and the quicker it is done the better for all concerned, except the grave digger....[65]

It seemed that everyone had a solution for the problem. Then, a major break in the Chambers Lake dam, located on the upper Poudre, resulted in flooding that took out the diversion structure shared by the Mercer Ditch and the water works supply canal. Costly repairs were undertaken to replace the "dam," leaving few additional dollars to address the filter problems. Simply cleaning the old sand and gravel filter in the reservoir and cleaning out the ditch were the only feasible, short-term solutions. High water caused by spring run-off caused the problem to return each year, and demands for improving the system continued.

The city nearly doubled in population in ten years and grew to 2,011 by 1890. Water mains were extended along several more streets. This taxed the already outdated water system even more. Colorado and the rest of the nation were heading into a major business depression, known as the Panic of 1893. A number of factors caused this depression, but many people placed blame on the repeal of legislation that provided for the purchase and coinage of silver. These federal actions resulted in the collapse of Colorado's silver boom. As the country returned to the gold standard, Fort Collins needed to find money to address the problems with its water system. Sand and silt in the supply canal reduced the efficiency of the turbines and pumps, and affected water quality. Ice in the supply ditch could all but cut off the city's water supply in winter. Regional droughts also restricted the source of power water for the turbines. By 1893, the pumps were in need of major mechanical overhauls, and the wooden penstock was rotting away.

A report was made to the council on November 20, 1893, about the conditions at the water works.

> The chairman of the Committee on Water Works reported that the supply ditch and filter needed to be cleaned, that the penstock needed caulking, that the wheels should be raised, the wing on the waste way should be repaired and the machinery needed a general overhauling, a coal house should be built and the dwelling needs a back porch. In fact, the works needed a great many repairs. On motion, the report was accepted.[66]

More Pipe. Archaeologists gained more evidence about pump house operations. A series of large ten- and twelve-inch pipes were located along the north wall leading into the 1894 addition, on the southeast exterior, and immediately in front of the main entrance. This ten-inch pipe, located along the north wall, may be associated with the 1894 steam engine room and/or the 1895 filtration room. The exterior pipe may have been connected to the pumps that carried pressurized water to Fort Collins. Courtesy, Richard Carrillo, Cuartelejo HP Associates.

In February 1894, a $228.50 contract was awarded to George Lynn, of Denver, to repair the pumps. The Water Works Committee, in April, also recommended to the council, that a steam pump and two new water wheels be purchased and that a stone wing wall, or dam, be built behind the pump house. This would increase the size and depth of the reservoir, and improve water pressure. In May, William Metcalfe was awarded the contract for the stonework, "… at $3.25 per perch [unit of measurement] for the stone work. This includes, also, all necessary evacuations, fillings, etc., and the tearing out of the old penstock and flume."[67] Six weeks later, another contract was made with Harry Davies to build an addition on to the north side of the pump house, for $597. (This is the second or center section of the present Water Works building.) The steam pump would allow water pressure to be maintained in times of low water and provide an alternative water pressure source. By early July, the *Weekly Courier* reported favorably on the progress of the work.

> A number of necessary changes and important repairs are being made to the city water works, at the pump house, which are designed to add to the durability of the works and to increase their efficiency. The old wooden penstock and flume, which had become badly decayed and therefore unsafe, have been torn out to give place to a solid stone retaining wall which is being laid up twenty-five feet high to hold the water in the settling reservoir above in check. This wall starts from bed rock and is six feet thick at the base and will be four feet thick at the top. It is being laid in Portland cement in the best and most substantial manner.
>
> A steel pipe, of sufficient capacity, will be placed in the wall at a suitable height to conduct water for driving the water wheels from the forebay above to a steel penstock below the wall, and from the penstock it will be applied to the wheels through other pipes. The water wheels were originally placed too low and had to labor continually in back water, which materially reduced their power and efficiency. When set again they will be raised above the level of the back water so that they may have a free and unimpeded discharge, thus giving them greater power for the work required

> in driving the pumps. The wheels, pumps, and all other machinery of the works will be overhauled and repaired and put into the best possible condition....
>
> While these changes and repairs are being made, a steam pump, having a capacity of 600 gallons per minute supplies the city with water, connection between the pump and water works canal having been made through a six-inch pipe extending to a point in the ditch above a temporary dam built across the canal a little above the head of the forebay. This pump is kept in motion day and night and at times Engineer [George] Giddings has had to crowd the discharge up to 675 gallons a minute in order to meet requirements. The pump will hereafter be a permanent fixture and kept ready for use at all times so that it can be put to work in case of accidents to and stoppage of the other pumps.[68]

However, replacement of the grossly inadequate filter in the bottom of the reservoir had to wait one more season. In May 1895, bids were advertised to put up a building over the filters. On June 1, a contract was signed with George W. King to construct this addition. (This is the third and most northerly section of the present Water Works building.) King was paid $400 in mid-July, possibly indicating that the work had been completed. The filters inside the new addition were sand and "perforated Aluminum Bronze plates." In mid-July 1896, high water caused many complaints about the quality of the water. The *Weekly Courier* came to the defense of the new filters, and explained their use and the hard job the two filter men had to do, in the July 16 edition of the paper.

> ....The capacity of the filters at the pumping works are not equal to the task of purifying the water as fast as it is needed. The water is first turned into the filters and leached through a bed of sharp sand twenty-six inches in thickness, and then strained through 1400 small

> strainers before it goes into the pipes. Ordinarily the sand is drenched and cleansed once an hour, then the water passing through it and the strainers is as clear as crystal, but when the river is muddy the sand has to be stirred up and cleansed every five or ten minutes, so that it is impossible to keep a sufficient supply of clear water running into the pipes to meet demand. On Wednesday the water was thick with mud and filth and the sand in the filters had to be drenched and cleaned every five minutes in order to get a supply through it. During this operation more or less muddy water unavoidably passes into the pipes and out of the hydrants. It is an affliction that must be borne for the present and until the city is able to provide a better and more perfect water system.[69]

The newspaper story's final statement indicated the problems with the original system's failure to keep up with the growing community's needs and the seasonal changes in the volume and quality of the Poudre's water. The filters had to be closed down when freezing weather occurred, and the old sand and gravel filter at the bottom of the reservoir had to be used. The filters were occasionally shut down when the water works engineer felt the filters could not adequately handle the volume of mud and silt caused by sudden, strong thunderstorms upriver. The *Weekly Courier* occasionally poked fun at city staff about these problems at the water works. On August 27, 1896, for example, "To Will Bell [water works engineer]—We are radically opposed to mixed drinks. Send us clear water. A mixture of ten per cent water and 90 per cent mud and manure doesn't make a first class table drink."[70] Major changes were in the future for Fort Collins' water supply.

The close of the nineteenth century must have had some climatic similarities to the end of the twentieth century. The *Weekly Courier* on April 4, 1897, printed a reminder to the townsfolk of an ordinance that had been passed in May 1894. The ordinance divided the town into two water districts—one west of College Avenue and the other east of that thoroughfare. Lawn watering, between May 1 and November 1, was restricted to three different days of the week for each district, with Sunday as a "no watering day." The restriction may have been put in place because of drought, or it may have been a result of the inadequacies of the water works system itself. The notice printed in the papers in August 1898, seems to support the latter explanation. "It shall be the duty of all water consumers, when the fire alarm is sounded, to at once turn off the water from their lawns and gardens...."[71] Another announcement, printed September 29, followed the same theme with new explanation: "It has been decided by the water committee that water cannot be used to water lawns after October 1. The water in the river is so low that it is impossible to get enough for domestic uses...."[72]

Two "tongue-in-cheek" quips in the *Weekly Courier* in April 1899, did a good job of summarizing water problems at the close of the century.

> The filter at the water works pumping station was put to work on Wednesday sifting the saw logs, railroad ties, tree limbs, toads, wrigglers, and the thickest mud out of the river water before it goes into the pipes....
>
> The river is carrying a big body of very dirty looking water these days, and it is only just getting its hand in, so to speak. When the floods come in May we may look for log cabins, bridges, rocks and whole sidehills to come floating down the stream.[73]

If Fort Collins wanted to move into the "modern" twentieth century looking like an "up-and-coming" Colorado city, the problems related to the water works system needed to be addressed and a solution found. The first decade of the new century was one of unprecedented growth as lambs and sugar beets created an agricultural boom. The town's population in 1900 was 3,053. By 1910, it grew at the phenomenal rate of 168.9 percent, to 8,210 souls. Many changes were afoot for the city.

Cache la Poudre at flood stage, 1904. Flood waters and other water quality issues between 1890 and 1903 forced Fort Collins residents to approve plans for a new water delivery system. Fort Collins could no longer claim that the original water works was a better facility than those of other Colorado towns. Courtesy, Fort Collins Public Library.

Chapter 5: Fort Collins Needs a Better Water Source

The new century dawned with the same concerns that had bedeviled the citizens of Fort Collins as the nineteenth century came to a close—how to provide the city with a clean, healthy domestic water supply. Typhoid fever, the turn-of-the-century scourge, did not spare Fort Collins. Local newspapers ran long articles with steps to take to prevent the ailment from making its way into the people's homes. Scott's Pharmacy even placed a newspaper advertisement headed, "DON'T DRINK WATER," as they advertised their, "full line of disinfectants and deodorants."[74] This was not a problem confined only to Fort Collins. The disease killed many Americans in the early twentieth century. Early aeronaut Wilbur Wright died of typhoid fever in 1912. City council attacked the source of some of the water supply's potential pollutants by passing an ordinance on December 17, 1900, banning the burial of "... night soil, grease trap, or cess pool materials ... in or on any premises within the city limits."[75] They also included fines for violations of the ordinance. Nevertheless, the problems of an inadequate water supply and filter system still needed to be addressed. (In 1888, Fort Collins completed its first sewer drainage system—a system of pipes that conveyed raw sewage to the Cache la Poudre River.)

Flooding, the already outdated filters, and drawing water for the old water plant from the river below Bellvue and part of Laporte, also with growing populations, continued to haunt the populace. Proposals to move the source of the water supply farther upriver were presented and then rejected during the next couple of years. Fort Collins, always a fiscally conservative community, had concerns about

Great Western Sugar Company, ca. 1920s. Replacement of Fort Collins first water delivery system became an even greater necessity in 1903 after local investors built the "sugar-beet factory" on East Vine Drive. Sugar-beet processing required large quantities of pure water, and the city supplied the factory's needs through a separate water main. Organized in 1902 as the Fort Collins Colorado Sugar Company, it consolidated with Great Western Sugar Company in 1904. The factory closed in 1955, and most of the associated structures were demolished in the late 1960s. Today, the Fort Collins Streets Department uses the former sugar warehouse at 625 Ninth St., as its office. Courtesy, Fort Collins Public Library.

Unloading sugar beets, 1904. In the late nineteenth century, scientists discovered that sugar beets grown in Northern Colorado tested high in sugar content. Local investors encouraged area farmers to raise the crop and began construction of Fort Collins Colorado Sugar Company's new plant. Completed in time for the 1903-1904 "campaign," farmers sold the perishable sugar beets to the factory, and in return, received beet pulp to use as livestock feed. Thus a by-product of the sugar mill fattened thousands of lambs each year in feedlots that surrounded Fort Collins in the early twentieth century. Courtesy, Colorado State University Photographic Services.

the original water bonds, which still had not been paid, and about passing a new bond issue to pay for a new water works. City Engineer William Rist presented his study for a new location and projected costs for pipelines, reported in the January 24, 1901, *Weekly Courier.* Rist strongly recommended a site nine miles from the city in the Poudre Canyon on the North Fork just above the point where the it flows into the main channel of the Cache la Poudre River. He received quotes for pipe from a few companies. The estimated cost for eighteen-inch cast iron pipe from the new site to the city was $175,395, and for eighteen-inch wooden stave pipe $81,376. City council proposed a bond election for early April for the new water works. The newspapers editorialized against the proposal, reminding the voters of the $105,000 in bonds left from the 1882 water works vote that would be added to the $185,000, giving the city a water-works debt of $290,000. The bond issue failed 233 to 41, so it was "back to drawing board" for city officials.

In 1902, selling the old water plant and granting a franchise for the construction and operation of

In Colorado, water is scarce whether or not there is a drought. A unique system of water allocation, often referred to as "first in time, first in right," states that the first person to divert water for beneficial use has the priority claim to that amount of water. This is extremely important in dry years when there may not be enough water for all users. Those with the earliest rights get the water first.

—Brian Werner,
"Irrigation Development in Northern Colorado:
A Brief History of How Water Influenced
the Development of the Fort Collins Region"
(Unpublished paper, Fort Collins Water
Works Archive, 2002).

The right to appropriate water had been provided by Colorado Territorial laws in 1861, but statutes for administering the priority of rights had not yet been established. By July 1874, upstream canals around Fort Collins were taking all water from the river, and the Greeley canals were dry. Union Colony leaders realized that some form of regulating authority was needed. That view was enhanced when downstream irrigators began to break down the diversion works of upstream canals.

A meeting of irrigators was held to consider the crisis. Greeley representatives argued for the principle of priority of water rights. No agreement was reached, but upstream water users promised to allow some water to flow down to Greeley's canals. Fortunately, heavy rains following the meeting made the test of that promise unnecessary. However, Union Colony was determined to get official recognition of the priority of water rights doctrine, and they succeeded two years later in getting it incorporated into the constitution of the new State of Colorado.

When in 1878 Benjamin Eaton began building the Larimer and Weld Canal to divert 571 cubic feet per second (usually the entire flow of the river) with its headgate upstream of all other canals, the unresolved issues of how to implement the prior rights doctrine again were raised by Cache la Poudre water users. The resulting state legislation of 1879 and 1881 were trailblazing steps added to the already landmark water law principles of beneficial use and priority in time, both credited to irrigators of the Cache la Poudre basin.

—Norman Evans,
"Colorado Irrigation History—
The Cache la Poudre River Basin"
(Unpublished paper, Fort Collins Water
Works Archive, 2002).

a water works was studied and rejected. The low priority of the city-owned water shares was another concern that was raised. The *Weekly Courier* published an article expressing this concern August 27.

> The city owns an appropriation of four cubic feet of water, which it purchased from A. T. Gilkinsin [Gilkison] several years ago, but senior appropriators have a right to 180 cubic feet before the city is entitled to a drop of its appropriation. The appropriations of the Pleasant Valley & Lake canal, of the Yeager ditch, the Jackson ditch, and the Taylor & Gill ditch all ante-date the city's appropriation, and if they all insist on their rights the city would not now be getting a drop. Exclusive of the water coming from other water sheds, the river is only carrying about sixty cubic feet per second, or about 120 feet less than the ditches named are entitled to before the city has a right to any under its appropriation.... Water Commissioner Hawley is kindly permitting just enough water to go through the city's headgates to supply the absolute necessities of the city consumers, ... This situation is likely to exist for the next week or ten days or until the water is no longer needed for irrigating crops.... Don't waste a drop unnecessarily....[76]

The spring of 1903 brought another vote on the new water works question, after the city council decided to "refund", or refinance, the old water bonds at a lower interest rate. The need for pure, safe water with enough pressure to fight fires was the paramount issue again. The result of the election was 139 in favor of the proposed installation and 53 against. By July 1, the council advertised for bids for the new water works. The *Weekly Courier* noted, "... [plans] embrace a complete gravity system with wooden stave pipe...."[77] Bids were opened in early August, and the contract for the pipeline was given to Holme and Allen for $123,778.18. The same firm received the contract to construct the "headworks, sedimentary basins and skimmers" for the new system. Work began December 10, 1903, and

New Fort Collins Water Works, n.d. This 1905 water facility met Fort Collins' increased demand for residential, commercial, and industrial uses for many more years. Now a part of Gateway Park on North Fork of the Cache la Poudre River, the new system relied on gravity flow to provide adequate water pressure in town. Courtesy, Fort Collins Public Library.

was completed exactly a year and a half later on June 10, 1905.

The new system had 13,400 feet of twenty-two-inch wooden pipe and 43,000 feet of eighteen-inch wooden pipe laid from the new water works to the city. The water mains were made of Oregon or Douglass fir with iron bands around the outside of the pipes. The new plant had a head of 425 feet above the city, creating a pressure of ninety-seven pounds per square inch. The total daily capacity of the new system was six million gallons. The new facility was a one-story brick building, thirty-seven by ninety-five feet. Once the new plant went into operation, the question of what to do with the old water works came before city council.

On June 28, 1905, the *Weekly Courier* ran a front-page article, "More Manufacturing For Fort Collins....City Council to Lease Pump House Water Power to G. H. Sethman." Sethman proposed using the old pump house to ".... manufacture transformers and other electrical appliances and to furnish light, power, and heat to patrons...."[78] The council agreed to lease the building for $1,200

Wooden pipe offered two major advantages: it was resistant to chemicals, and when soaked, it allowed water to flow with little resistance. Continuous stave wooden pipe was used to construct the 1905 water distribution system where water pressure was not a problem. Some of these staves were recycled as rafters in the 1930s when the chicken coop was under construction. Other staves were used to build the corral fence west of the barn. No wooden pipe was used in the 1882-1883 water delivery system.

—David Budge,
"Nineteenth-century Technology and the Water Works Site"
(Unpublished paper, Fort Collins Water
Works Archive, 2003).

a year with the proviso, or stipulation, that the city had the right to use the power if something happened to incapacitate the new water works system. In April of the next year, council agreed to lease the old water works supply ditch to the Larimer County Canal No. 2 Company for ninety-nine years. The ditch company then abandoned their old ditch that paralleled present-day Overland Trail. (It is interesting to note that the Fort Collins City Council will deal with that lease again in 2005.)

By August 1908, Fort Collins citizens were again agitating for a cleaner water supply. The "skimmers" at the new plant were only good for removing sediment from the water. Many people felt that a filter system also needed to be installed. The next summer, council decided to issue $75,000 worth of bonds to build a filter plant at the Poudre Canyon water works with a "Roberts Mechanical Filter" inside. The money raised also provided enough extra to build a covered storage basin, or reservoir, on Bingham Hill, where reserve filtered water supplemented demand in the city. The new "underground warehouse" was 150 feet wide, 250 feet long, and 25 feet tall. It increased the reserve water supply of the city "thirty fold," with its five million gallon capacity.[79] Contractors completed both projects in 1910.

Disposal of the equipment in the old water works first came up in an August 1916 newspaper article. Apparently, Sethman no longer was using it, if he ever did, since no mention of him appeared in the story. The article contained a good description of what was left in the old pump house.

> The pumping plant, which is located a mile south of Laporte and which formerly elevated the water into a standpipe directly west of the city, was erected in 1882 and 1883, the system being installed and put into use in 1883. It contains, in addition to a quantity of valves and pipes, a large boiler and engine and several pumps,...one of which is of a late model, and may bring more than its value as scrap.[80]

For some reason the scrapping was not carried out at that time. Perhaps the impending entry of the

United States into World War I stayed its execution, although scrap iron would have been a valuable commodity during that time. The annual report of the Public Works Department issued in 1922 by Frank Goeder, Commissioner of Public Works, and John Revell, City Engineer, summarized the previous five years' work of that department. The report told of the city's efforts in 1920.

> The machinery at the old pump house near Laporte was wrecked. Many of the valves, much lumber and some reinforcing steel was salvaged and saved for various uses in the water department. Twenty-one and five-tenths tons of scrap iron was sold to the Giddings Manufacturing Company at $15.50 per ton, netting the city $333.20. The old boiler and smoke stack were sold for $225.00 and in addition to the above, there was much other material which probably resulted in a net gain to the city of more than one thousand dollars from the wrecking of the obsolete equipment.[81]

During the next half century, the old water works building had various other municipal purposes. It contained a city mechanical shop for many years and became a storehouse for city equipment. The city's Historic Landmarks Preservation Commission designated it as a local landmark in 1971. The on-site residence housed a city employee and his family, and assured some protection for the site until the early 1980s. But minimal maintenance on the structure led to severe deterioration.

Jim and Doris Greenacre organized a small group of preservationists in the late 1980s, and convinced the city, in 1988, to support a structural stabilization project. In 1999, the Water Works property was listed on the State Register of Historic Places. The Poudre Landmarks Foundation initiated another effort in the late 1990s to preserve the building and convert it into an interpretive center for the history of municipal and agricultural water use in the area. The Friends of the Water Works and the Foundation continue to work toward that goal.

Endnotes

1 Ansel Watrous, *History of Larimer County, Colorado* (Fort Collins, Colorado: Courier Printing Co., 1911. Reprint. Fort Collins, Colorado: Miller Manor Publications, 1972), 44.
2 *History of Larimer County,* 26-27.
3 *History of Larimer County,* 246.
4 *History of Larimer County,* 242.
5 "Water Works For Collins," *Fort Collins Weekly Express* (March 26, 1880): 2.
6 Ibid.
7 "Better Water Supply Needed," *FCWE* (December 30, 1880): 2.
8 "A Water Works Scheme," *FCWE* (January 27, 1881): 2.
9 "Another Plan For Water Works," *FCWE* (February 10, 1881): 1.
10 "Editorial," *FCWE* (May 12, 1881): 2.
11 "Editorial," *FCWE* (August 18, 1881): 4.
12 "City News," *Fort Collins Courier* (January 19, 1882): 4.
13 "The water works project has fallen through... ," *FCWE* (May 25, 1882): 1.
14 Ibid.
15 "If any person... ," *FCC* (June 1, 1882): 2.
16 "Editorial Notes," *FCWE Supplement* (June 1, 1882): 1.
17 "The City Hall," *FCWE* (June 7, 1882): 3.
18 "The Hook and Ladders meet for practice," *FCC* (June 29, 1882): 3.
19 "Furious Flames," *FCC* (September 16, 1882): 2.
20 Ibid.
21 Ibid.
22 "Notes," *FCC* (September 16, 1882): 2.
23 "The majority rules... ," *FCWE* (September 21, 1882): 2.
24 "Engineers take measurements up-river... ," *FCWE* (September 27, 1882): 3.
25 "Water Works," *FCWE* (October 4, 1882): 2.
26 "Proposal for Bids," *Fort Collins Daily Express* (October 4, 1882): 4.
27 "Opening Bids," *FCDE* (October 30,1882): 2.
28 "Bids Wanted," *FCDE* (November 2, 1882): 4.
29 "The contract... ," *FCDE* (November 13, 1882): 4.
30 "Home News," *FCDE* (November 20, 1882): 4.
31 "City Engineer Handy ... ," *FCDE* (November 25, 1882): 4.
32 "The Town Board," *FCDE* (December 5, 1882): 4.
33 "Mr. E. S. Alexander, . . ," *FCDE* (December 6, 1882): 4.
34 "City Water Works," *FCDE* (December 18, 1882): 1.
35 "Eight car loads... ," *FCDE* (January 23, 1883): 4.
36 "J. J. Bradley... ," *FCDE* (January 26, 1883): 4.
37 "Water Service... ," *FCDE* (March 22, 1883): 4.

38 "Over $60,000… ," *FCDE* (March 31, 1883): 2.
39 "The machinery… ," *FCDE* (April 10, 1883): 4.
40 "The city water bonds… ," *FCDE* (April 16, 1883): 4.
41 "Mayor Emigh… ," *FCDE* (April 17, 1883): 1.
42 Ibid.
43 "Aqua Pura," *FCC* (April 19, 1883): 1.
44 "J. G. Lunn… , *FCDE* (April 25, 1883): 4.
45 "Work on… ," *FCDE* (April 26, 1883): 4.
46 "The water works… ," *FCDE* (May 24, 1883): 4.
47 "Testing… ," *FCDE* (May 31, 1883): 1.
48 Ibid.
49 "The Water Works Finished… ," *FCDE* (June 1, 1983): 1.
50 "The water works pumps…. ," *FCDE* (June 5, 1883): 2.
51 Ibid.
52 "Now We Can Crow… ," *Fort Collins Daily Evening Courier* (June 7, 1883): 1.
53 Ibid.
54 "Farewell Feast," *FCDE* (June 8, 1883): 1.
55 Ibid.
56 "Formally Accepted," *FCDE* (June 9, 1883): 1.
57 "The first water… ," *FCDE* (June 9, 1883): 4.
58 The Works In Operation," *FCDE* (June 12, 1883): 1.
59 "Over sixty… ," *FCDE* (August 15, 1883): 2.
60 "Consumers of water… ," *FCDE* (August 23, 1883): 4.
61 "Water Superintendent Owens… ," *FCWE* (February 7, 1885): 1.
62 "Fort Collins has… ," *FCWE* (December 31, 1887): 1.
63 "Municipal Improvements," *Fort Collins Weekly Courier* (January 9, 1890): 1.
64 "Citizens Meet the Board of Health," *FCWE* (May 30, 1891): 1.
65 "The time is near… ," *FCWC* (June 18, 1891): 4.
66 "Fort Collins City Council Minutes," *Book D* (November 20, 1893): 77.
67 "The contract … ," *FCWC* (May 10, 1894): 1.
68 "Water Works Undergoing Repairs," *FCWC* (July 5, 1894): 1.
69 "The river, . . ," *FCWC* (July 16, 1896): 1.
70 "To Will Bell… ," *FCWC* (August 27, 1896): 1.
71 "Notice to Water Consumers," *FCWC* (August 4, 1898): 1.
72 "Notice to Water Users," *FCWC* (September 29, 1898): 1.
73 "The filter… ," *FCWC* (April 20, 1899): 1.
74 "Don't Drink Water," *FCWC* (December 20, 1900): 1.
75 "Ordinance No. 6-1900," *FCWC* (December 20, 1900): 2.
76 "City's Water Supply May Be Cut Off," *FCWC* (August 27, 1902): 2.
77 "The New Water Works," *FCWC* (July 1, 1903): 11.
78 "More Manufacturing For Fort Collins," *FCWC* (June 28, 1905): 1.
79 "City Council Decides On Roberts Mechanical Filter," *Fort Collins Evening Courier* (July 20, 1909): 1.
80 "Old Water Plant Is To Be Junked," *FCEC* (August 17, 1916): 9.
81 Frank P. Goeder and John Revell, "Water Department," *Report of Commissioner of Works and City Engineer, City of Fort Collins, Colorado* (1922): 10.

Suggested Readings

Duncan, C. A. *Memories of Early Days in the Cache La Poudre Valley.* Timnath, Colorado: The Columbine Club of Timnath, n.d.

Case, Stanley R. *The Poudre: A Photo History.* Bellvue, Colorado: Privately Published, 1995.

Evans, Howard E. and Mary A. Evans. *Cache La Poudre: The Natural History of a Rocky Mountain River.* Niwot, Colorado: University Press of Colorado, 1991.

Fry, Norman W. *Cache La Poudre "The River": As Seen From 1899.* Privately Published, n.d.

Gray, John S. "The River Acquires Its Name," *The Poudre River Magazine.* Denver: Gro-Pub Group, 1976, 10-12.

Greeley, Horace. *An Overland Journey from New York to San Francisco in the Summer of 1859.* New York: C. M. Saxon, Barker, & Co., 1860.

Nortier, Molly and Michael Smith. *"From Bucket to Basin": 100 Years of Water Service.* Fort Collins, Colorado: City of Fort Collins, 1982.

Steinel, Alvin T. *History of Agriculture in Colorado, 1858 to1926.* Fort Collins, Colorado: The State Board of Agriculture, 1926.

Swanson, Evadene B. *Fort Collins Yesterdays.* Fort Collins, Colorado: Privately Published, 1976.

Tremblay, William. *The June Rise: The Apocryphal Letters of Joseph Antoine Janis.* Logan, Utah: Utah State University Press, 1994.

Tyler, Daniel. *The Last Water Hole in the West: The Colorado-Big Thompson Project and the Northern Colorado Water Conservancy District.* Niwot, Colorado: University Press of Colorado, 1992.

Waddell, Karen and Erin Christensen. *How the Waste Was Won: A Century of Wastewater Service in Fort Collins, 1888–1988.* Fort Collins, Colorado: City of Fort Collins, 1988.

Watrous, Ansel. *History of Larimer County, Colorado.* Fort Collins, Colorado: Courier Printing Co., 1911. Reprint. Fort Collins, Colorado: Miller Manor Publications, 1972.

Whiteley, Lee. *The Cherokee Trail: Bent's Old Fort to Fort Bridger.* Privately Published, 1999.

Bibliography

Barnes, Albert. "Land Surveying for the Fort Collins Waterworks." Unpublished paper. Fort Collins Water Works Archive, 2003.

Budge, David. "Nineteenth-century Technology and the Water Works Site." Unpublished paper. Fort Collins Water Works Archive, 2003.

Carrillo, Richard. "Historical Archaeology and the Fort Collins Water Works (5LR749)." Unpublished draft report. Fort Collins Water Works Archive, 2003.

Cook, Kevin. "Wildlife of the 1882-1883 Fort Collins Water Works." Unpublished paper. Fort Collins Water Works Archive, 2002.

Dion, Sandy. *1882 Fort Collins Water Works: Newspaper Research.* 2 vols. Fort Collins, Colorado: Poudre Landmarks Foundation, 1999.

Evans, Norman. "Colorado Irrigation History—The Cache la Poudre River Basin." Unpublished paper. Fort Collins Water Works Archive, 2002.

"Fort Collins City Council Minutes." *Book D.* Fort Collins City Clerk's Office.

Fort Collins Daily Evening Courier, 1883-1898. Microfilm, Fort Collins Public Library.

Fort Collins Daily Express, 1881-1884. Microfilm, Fort Collins Public Library.

Fort Collins Evening Courier, 1902-1912. Microfilm, Fort Collins Public Library.

Fort Collins (weekly) Courier, 1880-1920. Microfilm, Fort Collins Public Library.

Fort Collins Weekly Express, 1880-1916. Microfilm, Fort Collins Public Library.

Goeder, Frank P. and John Revell. "Water Department," *Report of Commissioner of Works and City Engineer, City of Fort Collins, Colorado,* 1922.

"History of the Fort Collins Water Works, July 1920." Unpublished manuscript. Fort Collins Public Library Local History Archive and Fort Collins Water Works Archive.

Hoefer, Philip. "Trees of the Old Water Works." Unpublished paper. Fort Collins Water Works Archive, 2002.

Quinnell, Susan. "Historical Interpretation Report of the Building Materials of the Fort Collins Water Works Building, 2005 North Overland Trail." Unpublished paper. Fort Collins Water Works Archive, 2001.

Watrous, Ansel. *History of Larimer County, Colorado.* Fort Collins, Colorado: Courier Printing Co., 1911. Reprint. Fort Collins, Colorado: Miller Manor Publications, 1972.

Werner, Brian. "Irrigation Development in Northern Colorado: A Brief History of How Water Influenced the Development of the Fort Collins Region." Unpublished paper. Fort Collins Water Works Archive, 2002.

Acknowledgments

The Friends of the Water Works and the Poudre Landmarks Foundation wish to thank all who assisted with the creation of this book. Our special thanks goes to local historian, Wayne Sundberg, for writing the bulk of Fort Collins' First Water Works. And, we want to recognize the contributors who wrote manuscripts about topics that pertain to the 1882-1883 Fort Collins Water Works and surrounding acreage: Albert H. Barnes, David Budge, Richard Carrillo, Kevin Cook, Norman Evans, Philip Hoefer, Susan Quinnell, and Brian Werner. In addition to offering your specialized knowledge for use in this publication, your work will be a resource for future site interpretation.

The Friends editorial board composed of Jacques Rieux, Norman Evans, and David Budge met with me to provide overall direction for the project, and several dedicated members of the Friends served as careful readers of the manuscript, not just once but often several times. Tom Boardman, David Budge, Jane Hail, Gayla Johnson, Jean Petersen, Jacques Rieux, and Wayne Sundberg provided this invaluable service and reduced the probability of error.

We also thank Gheda Gayou, Historic Preservation Specialist, State Historical Fund, and Modupe Labode, Chief Historian, Colorado Historical Society, for their insightful comments. Of course, this publication would not be possible without the financial assistance of the State Historical Fund and the Colorado Historical Society. And, thanks to Friends member Sandy Dion, who prepared and submitted the grant proposal for this project in 2001, and for her valuable newspaper research, which adds so much to the text.

Generous help in locating the photographs and maps to illustrate the text came from Rheba Massey, Fort Collins Public Library; Barbara Dey, Colorado Historical Society; Richard Carrillo, Cuartelejo HP Associates; Susan Quinnell; Tom Boardman; Wayne Sundberg; David Budge; and Lisa Steffes.

Finally, we wish to recognize the careful work of book designer Gail Blinde, who prepared the pages for Citizen Printing's press. All helped us realize our long-anticipated dream of publishing this book about Fort Collins' First Water Works.

Susan Hoskinson,
Editor

Index